Methods to Determine Enzymatic Activity

Editors

Alane Beatriz Vermelho

Biotechnology Center- BIOINOVAR
Bioenergy Biocatalysis and Bioproducts, and
Institute of Microbiology Paulo de Góes
Federal University of Rio de Janeiro
Rio de Janeiro
Brazil

&

Sonia Couri

Federal Institute of Education
Science and Technology of Rio de Janeiro
Rio de Janeiro
Brazil

CONTENTS

FOREWORD

Enzymes play a crucial role in all living organisms, being the best chemists that nature has "invented" (from the evolutionary viewpoint, of course). In the last decades these catalysts started to be used more and more in industrial, biotechnological processes, due to the fact that all chemical processes which normally need extreme conditions (high temperature and/or pressure, extremely acidic or alkaline media, *etc.*) can be done in much milder conditions, at temperatures of 25-40 °C, almost neutral pH, and in water as solvent. Considering the huge number of enzymes known so far and the new ones which are constantly being discovered and characterized in organisms all over the phylogenetic tree (e.g., extremophiles, Archaea, *etc*), it is no wonder that biotechnologies that use them extensively have grown exponentially in the last period.

In this context, I was delighted to read "Methods to Determine Enzymatic Activity", edited by Alane Beatriz Vermelho and Sonia Couri, which comprises a nice collection of 13 reviews, all of them from Brazilian scientists, dealing with these topics. Each chapter presents in a very nice manner the main reaction(s) catalyzed by the considered enzyme, its sources, purification, characteristics, followed by the detailed description of the assay methods used to determine the activity (as well as inhibition/activation) of these enzymes. Many representatives are taken into consideration, such as pectinases, peroxidases, enzymes with chitinolytic activity, cellulases, amylases, xylanases, lipases, phenoloxidases, transglutaminases, keratinases, peptidases (mainly serine and metallo-proteinases are considered), tannases and ureases. Most of these enzymes have important applications in the food, textile, leather, biofuel production, pharmaceutical, cosmetics, fine chemicals, biomaterials, paper/cellulose, and detergent industries.

The Editors and the authors did an excellent work in presenting in an exhaustive and highly professional way the state-of-the-art in all these field. I warmly recommend this eBook to students, researchers and scientists from diverse fields, as due to the simplicity in which the material is presented, they will be able to resolve in a quick way some research problems related to assay methods of enzymes with many applications in biotechnology.

Claudiu T. Supuran

University of Florence
Italy

PREFACE

The aim of this book is to provide an updated revision of the most important and simple methods to analyze enzymatic activity. Qualitative and quantitative methods for the industrial enzymes are described.

Microbial enzymes are increasingly replacing conventional chemical catalysts in a range of industrial processes in industrial since the process of biocatalys is ecologically correct. Moreover, the conditions to obtain and optimize the production of enzymes are easily controlled in bioreactors and the microorganisms can also be manipulated genetically to improve the desirable characteristics of a biocatalyzer. The main industries that apply microbial enzymes are: the food, textile, leather, biofuel pharmaceutical, cosmetics, fine chemicals, biomaterials, paper and cellulose and detergent industries. In this context a comprehensive overview of the methods used for enzyme detection would be useful. The book contains 13 chapters and presents the main methods to analyze the activity of pectinases, peroxidases, chitinases, cellulases, amylases, xylanases, lipases, phenoloxidase, transglutaminases, keratinases, peptidases, tannases and ureases.

The 1st chapter of this book is about the pectinases, one of the enzymes presently gaining much attention in the fruit and textile industries. This chapter describes the current state of knowledge of the main assays used to quantify pectinolytic enzymes and their activity. The second chapter focuses on the peroxidases. These enzymes are ubiquitous enzymes that catalyze the oxidation of lignin and other phenolic compounds at the expense of hydrogen peroxide (H_2O_2) in the presence of a mediator. They have been classified into many types based on their source and activity. One of the major application of the peroxidases is in the bioremediation processes.

In the 3th and 4th chapter, the chitinases and cellulases are discussed. Chitinolytic enzymes are gaining importance due to their biotechnological applications. Particularly, chitinases are used in agriculture to control plant pathogens.The main methods for the detection and measurement of chitinolytic activity will be presented. Cellulase has been used in the bioconversion process of lignocellulosic materials into bioethanol and biobased products, within the biorefinery concept. Also, cellulases have been widely used in various industries such as the textile, food, beer and wine, animal feed, pulp and paper industries. This chapter describes the current state of knowledge on assays for the glycoside hydrolase complex.

The 5th chapter describes enzymes the detection methods for all types of amylase, a group of important industrial enzymes with applications in food production, and the production of sugars from starch.

Xylanase and lipases detection methods are respectively in the 6th and 7th chapter. Xylanases play a vital role in depolymerizing xylan, the major component of hemicellulose. These enzymes have been used in traditional fields such as food, animal food and paper industries. The use of xylanase in producing sugars and other chemicals from lignocelluloses in recent years is increasing. Lipases hydrolyze triacylglycerols to fatty acids, diacylglycerols, monoacylglycerols and glycerol and under certain conditions, catalyze reverse reactions such as esterification and transesterification. They are used in various industries such as: the food, waste water treatment, cosmetics, biodiesel, pharmaceutical and detergent industries, as well as in biofuels.

The 8th chapter is about phenoloxidase, which are copper enzymes that catalyze the oxidation of phenolic compounds. Transglutaminases are focused in chapter 9[th.] They are biological "glues" capable of catalyzing acyl transfer reactions by introducing covalent cross-links between proteins, as well as peptides and various primary amines. Their most important use is in the food industry.

Qualitative and quantitative methods for peptidases and keratinases are described in the chapters 10 and 11 respectively. The proteolytic enzymes account for 25% of the industrial enzyme market and present multiple applications including the food and detergent industry, cosmetic, animal food, leather depilation, textile Pharmaceutical sectors.

Lastly, the chapters 12 and 13 report the major methods used for tannases and ureases. Tannases are acyl hydrolases that act on tannins. The major uses are in the production of gallic acid and glucose, beer processing and clarification of juices and the ureases catalyze the hydrolysis of urea into ammonia and carbon dioxide. They are mainly important in agriculture, medicine and clinical analysis.

We would like to thank the authors for their excellent work, their dedication and updated information.

Alane Beatriz Vermelho

Institute of Microbiology Paulo de Góes
Biotechnology Center- Bioinovar
Federal University of Rio de Janeiro
Rio de Janeiro
Brazil

&

Sonia Couri

Federal Institute of Education
Science and Technology of Rio de Janeiro
Rio de Janeiro
Brazil

List of Contributors

Adriana M. Fróes

Department of General Microbiology, Institute of Microbiology Paulo de Góes, Federal University of Rio de Janeiro, Rio de Janeiro, Brazil

Alane Beatriz Vermelho

Department of General Microbiology, Institute of Microbiology Paulo de Góes-Biotechnology Center- Bioinovar, Federal University of Rio de Janeiro, Rio de Janeiro, Brazil

Aline Machado de Castro

Biotechnology Division, Research and Development Center, PETROBRAS, Rio de Janeiro, Brazil

Ana Maria Mazotto

Department of General Microbiology, Institute of Microbiology Paulo de Góes, Federal University of Rio de Janeiro, Rio de Janeiro, Brazil

Andrea M. Salgado

Departament of Biochemical Engineering, Laboratory of Biological Sensors, School of Chemistry, Federal University of Rio de Janeiro, Rio de Janeiro, Brazil

Antonio Carlos Augusto da Costa

Department of Biochemical Process Technology, Institute of Chemistry, State University of Rio de Janeiro, Rio de Janeiro, Brazil

Ariana F. Melo

Departament of Biochemical Engineering, Laboratory of Biological Sensors, School of Chemistry, Federal University of Rio de Janeiro, Rio de Janeiro, Brazil

Bernardo Dias Ribeiro

School of Chemistry, Federal University of Rio de Janeiro, Rio de Janeiro, Brazil

Cristiane Sanchez Farinas

Embrapa Instrumentation, Rio de Janeiro, Brazil

Daniel Paiva

Institute of Microbiology Paulo de Góes, Federal University of Rio de Janeiro, Rio de Janeiro, Brazil

Edilma Paraguai de Souza

Institute of Microbiology Paulo de Góes, Federal University of Rio de Janeiro, Rio de Janeiro, Brazil

Giseli Capaci Rodrigues

Institute of Microbiology Paulo de Góes, Federal University of Rio de Janeiro, Rio de Janeiro, Brazil

Juliana P. Rosa

Department of General Microbiology, Institute of Microbiology Paulo de Góes, Federal University of Rio de Janeiro, Rio de Janeiro, Brazil

Lívia Maria da C. Silva

Departament of Biochemical Engineering, Laboratory of Biological Sensors, School of Chemistry, Federal University of Rio de Janeiro, Rio de Janeiro, Brazil

Maria Alice Zarur Coelho

Departament of Biochemical Engineering, Biological Systems Engineering, School of Chemistry, Federal University of Rio de Janeiro, Rio de Janeiro, Brazil

Mônica Caramez Triches Damaso

Embrapa Agroenergy, Brasília, Brazil

Mônica P. Gravina-Oliveira

Institute of Microbiology Paulo de Góes, Federal University of Rio de Janeiro, Rio de Janeiro, Brazil

Priscilla F. F. Amaral

School of Chemistry, Federal University of Rio de Janeiro, Rio de Janeiro, Brazil

Rodrigo F. Souza

Department of General Microbiology, Institute of Microbiology Paulo de Góes, Federal University of Rio de Janeiro, Rio de Janeiro, Brazil

Rodrigo P. Nascimento

School of Chemistry, Federal University of Rio de Janeiro, Rio de Janeiro, Brazil

Romulo Cardoso Valadão

Federal Rural University of Rio de Janeiro, Institute of Technology, Seropédica, Rio de Janeiro, Brazil

Rosalie R. R. Coelho

Department of General Microbiology, Institute of Microbiology Paulo de Góes, Federal University of Rio de Janeiro, Rio de Janeiro, Brazil

Sabrina Martins Lage Cedrola

Department of General Microbiology, Institute of Microbiology Paulo de Góes, Federal University of Rio de Janeiro, Rio de Janeiro, Brazil

Selma da Costa Terzi

Embrapa Food Tecnology, Rio de Janeiro, Brazil

Sonia Couri

Federal Institute of Education, Science and Technology of Rio de Janeiro, Rio de Janeiro, Brazil

Thaís Fabiana Chan Salum

Embrapa Agroenergy, Brasília, Brazil

ACKNOWLEDGEMENTS

This work was supported by the Coordenação de Aperfeiçoamento de Pessoal de Nível Superior (CAPES), Fundação Carlos Chagas Filho de Amparo à Pesquisa do Estado do Rio de Janeiro (FAPERJ), Conselho Nacional de Desenvolvimento Científico e Tecnológico (MCT/CNPq) and Centro de Pesquisas Leopoldo Américo Miguez de Mello (CENPES) PETROBRAS.

2

Send Orders for Reprints to reprints@benthamscince.net

CHAPTER 1

Methods of Determination of Pectinase Activity

Sonia Couri[1,*], Selma da Costa Terzi[2], Bernardo Dias Ribeiro[3], Antonio Carlos Augusto da Costa[3] and Cristiane Sanchez Farinas[4]

[1]*Federal Institute of Education, Science and Technology of Rio de Janeiro, Rio de Janeiro, Brazil;* [2]*Embrapa Food Tecnology, Rio de Janeiro, Brazil,* [3]*School of Chemistry, Federal University of Rio de Janeiro, Rio de Janeiro, Brazil and* [4]*Embrapa Instrumentation, Rio de Janeiro, Brazil*

Abstract: Pectinases constitute a group of enzymes widely used by the fruit and textile industries. Besides these industrial applications, these enzymes play an important biological role in protoplast fusion technology and plant pathology. Since the applications of pectinases in various fields are widening are becoming wider, it is important to understand the nature and properties of these enzymes for an efficient and effective application. For the past few years, intensive research has been carried out on isolation and characterization of pectinases. They are a group of enzymes that contribute to the degradation of pectin by various mechanisms and can be classified as esterases, eliminative depolymerases (lyases) and hydrolytic depolymerases (polygalacturonases). This chapter will describe the current state of knowledge of the main assays used to quantify pectinolytic enzymes and their activity.

Keywords: Enzyme, pectinase, esterases, depolymerases, galacturonic acid, pectic substances, protopectin, pectic acid, dinitrosalicylic acid.

PECTIC SUBSTANCES

Pectic substances, present in middle lamella and primary cell wall of higher plants, contribute for firmness and structure of the plant tissues. They are high molecular weight, negatively charged, acidic, complex glycosidic macromolecules (polysaccharides) that are present in the Plant Kingdom. They have a great impact on the viscosity and turbidity of fruit juices and is also well known for their ability to form gels.

Chemically, pectic substances are essentially branched heteropolysaccharides containing from a few hundreds to a thousand building-blocks per molecule. They consist of galacturonans and rhamnogalacturonans in which the C-6 carbon of

---

*Address correspondence to Sonia Couri: Federal Institute of Education, Science and Technology of Rio de Janeiro, Brazil; Tel: +55 (21) 2566-7733; E-mail: sonia.couri@gmail.com.

galactate is oxidized to a carboxyl group, the arabinans and the arabinogalactans. These substances are a group of complex colloidal polymeric materials, composed largely of a backbone of anhydrogalacturonic acid units. The carboxyl groups of galacturonic acid are partially esterified by methyl groups and partially or completely neutralized by sodium, potassium or ammonium ions. Some of the hydroxyl groups on C2 and C3 may be acetylated. The primary chain consists of α-D-galacturonate units α-(1→4) linked, with 2–4% of L-rhamnose units β-(1→2) and β-(1→ 4) linked to the galacturonate units (Fig. **1**) [1, 2].

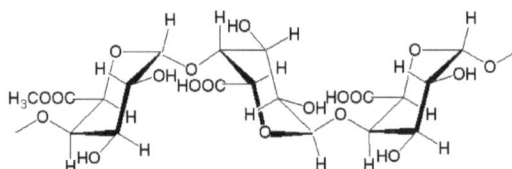

Figure 1: Primary structure of pectic substances (Source: Adapted from Rexova-Benkova and Markovic, 1976) [3].

Differently from proteins, lipids and nucleic acids, as a polysaccharide, pectic substances do not have a defined molecular weight. The relative molecular mass of pectic substances ranges from 25 to 360 kDa.

The American Chemical Society classified pectic substances into four main types as follows [4]:

- Protopectin is the water insoluble pectic substance present in the intact tissue. Protopectin on restricted hydrolysis yields pectin or pectic acids;

- Pectic acid is the soluble polymer of galacturonans that contains negligible amounts of methoxyl groups Normal or acid salts of pectic acid are called pectates;

- Pectinic acids is the polygalacturonan chain that contains from 0 to 75% >0 and <75% methylated galacturonate units. Normal or acid salts of pectinic acid are referred to as pectinates, and;

- Pectin (polymethyl galacturonate) is the polymeric material in which, at least 75% of the carboxyl groups of the galacturonate units are esterified with methanol. It provides rigidity on cell wall when it is bound to cellulose in the cell wall.

Pectinases

Pectinase is a generic name for a family of enzymes that catalyze hydrolysis of the glycosidic bonds in the pectic polymers. Pectinases were some of the first enzymes to be used in industry. Their commercial application started in the 30's for the preparation of wines and fruit juices. But it was only in the 60's that the chemical nature of plant tissues became known, and with this knowledge, scientists began to use enzymes more efficiently. As a result, pectinases are today one of the enzymes commonly used by the commercial sector [5].

Pectinases are produced during the natural ripening process of some fruits, where together with cellulases, they help to soften cell walls. These enzymes are also secreted by plant pathogens such as the fungus *Monilinia fructigena* and the soft-rot bacterium *Erwinia carotovora*, as part of their strategy for penetrating the plant host cell walls. In fact, the products of such enzyme (oligosaccharins) act as a signal which induces uninfected cells to protect themselves.

It has been reported that microbial pectinases account for 25% of the global food enzymes sales. Almost all the commercial preparations of pectinases are produced from fungal sources [2].

The major commercial source of pectinases is the filamentous fungus *Aspergillus niger*. They are most widely used in industries because this strain posses GRAS (Generally Regarded as Safe) status [6] so that metabolites produced by this strain can be safely used. These fungi produce a mixture of enzymes, which, along with others such as cellulases, are widely used in the fruit juice industry where they are used to help extract, clarify and modify fruit juices. Enzymes in the pectinase group include polygalacturonases, pectin methyl esterase and pectin lyases [7].

Pectinases are mainly used for increasing filtration efficiency, clarification of fruit juices and are widely used in maceration, liquefaction and extraction of vegetable tissues. They help to decrease viscosity and prevent gelation in the extracts. Therefore pectinase enzymes are commonly used in processes involving the degradation of plant materials, such as speeding up the extraction of fruit juice from fruit, including apples. Pectinases have also been used in wine production since the 1960s. Recently pectinases have been also used in the paper and pulp industry in addition to cellulases [8].

Pectinases are classified under three headings according to the following criteria: (a) whether pectin, pectic acid or oligo-D-galacturonate is the preferred substrate; (b) whether pectinases act by trans-elimination or hydrolysis; and, (c) whether the cleavage is random (endo-, liquefying of depolymerizing enzymes) or endwise (exo- or saccharifying enzymes). The three major types of pectinases are as follows [5]:

i.   Protopectinases: degrade the insoluble protopectin and give rise to highly polymerized soluble pectin;

ii. Esterases: catalyze the de-esterification of pectin by the removal of methoxy esters; and;

iii. Depolymerases: catalyze the hydrolytic cleavage of the α-(1$\rightarrow$ 4)-glycosidic bonds in the D-galacturonic acid moieties of the pectic substances.

Different pectic substances and their mode of reaction are illustrated in Fig. **2** [9, 10].

Depolymerases act on pectic substances by two different mechanisms: 1) hydrolysis, in which they catalyze the hydrolytic cleavage with the introduction of water across the oxygen bridge; and, 2) trans-elimination lysis, in which they break the glycosidic bond by a trans-elimination reaction without any participation of water molecule [10, 11].

Depolymerases can be subdivided into four different categories, depending on the preference of enzyme for the substrate, the mechanism of cleavage and the splitting of the glycosidic bonds [3].

Polygalacturonase and polymethylgalacturonase breakdown pectate and pectin, respectively by the mechanism of hydrolysis. However, polygalacturonate lyase and polymethylgalacturonate lyase breakdown pectate and pectin by b-elimination respectively.

Pectinesterases also known as pectinmethyl hydrolase, catalyzes deesterification of the methoxyl group of pectin forming pectic acid. The enzyme acts preferentially on a methyl ester group of galacturonate unit next to a non-esterified galacturonate unit.

a)

PM/PG

b)

PE

PE

c)

PL/PGL

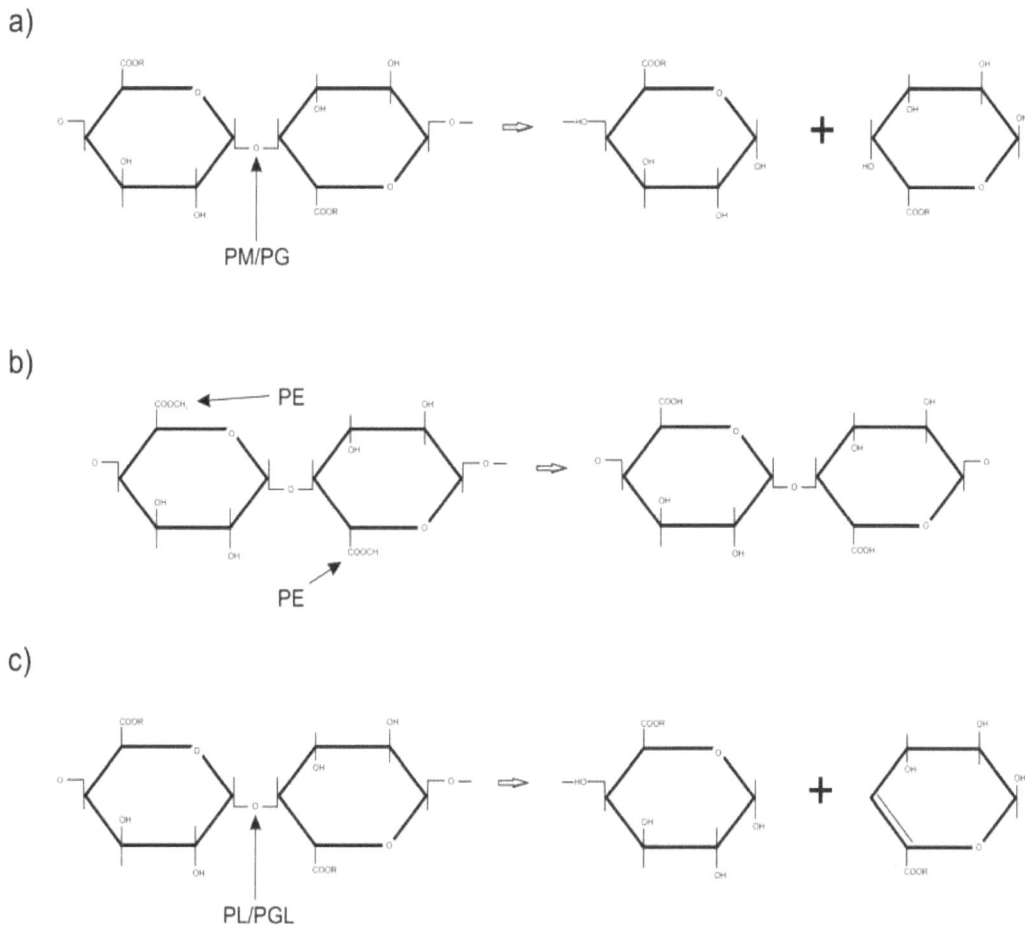

Figure 2: Mode of action of pectinases: (a) R = H for PG and CH3 for PMG; (b) PE; and (c) R = H for PGL and CH3 for PL. The arrow indicates the place where the pectinase reacts with the pectic substances. PMG, polymethylgalacturonases; PG, polygalacturonases (EC 3.2.1.15); PE, pectinesterase (EC 3.1.1.11); PL, pectin lyase (EC-4.2.2.10). Source: Adapted from Sathyanarayana and Panda, 2003 [8].

Depending upon the mode of action, *i.e.,* random or terminal, these enzymes are termed as Endo or Exo enzymes, respectively (Fig. **2**).

In practice, however, enzyme activity is affected by many things (*e.g.,* pH, temperature, the presence of inhibitors or cofactors) which will affect the results to be obtained. Pectinases have an optimum temperature and pH at which they are most active. For example, a commercial pectinase might typically be activated at 45 to 55 °C and work well at a pH of 4.8 to 5.0.

Table 1: Classification of pectinolytic enzymes (Source: Adapted from Ranveer et, al, 2005)

Enzyme	E.C. no.	Modified EC systematic name	Action mechanism	Action pattern	Primary substrate	Product
Esterase						
1. Pectin methyl esterase	3.1.1.11		Hydrolysis	Random	Pectin	Pectic acid + methanol
Depolymerizing enzymes						
a. Hydrolases						
1. Protopectinases			Hydrolysis	Random	Protopectin	Pectin
2. Endopolygalacturonase	3.2.1.15	Poly-(1-4)-α-D-galactosiduronate lyase.	Hydrolysis	Random	Pectic acid	Oligogalacturonates
3. Exopolygalacturonase	3.2.1.67	Poly-(1-4)-α-D-galactosiduronate glycanohydrolase	Hydrolysis	Terminal	Pectic acid	Monogalacturonates
4. Exopolygalacturonan-digalacturono hydrolase.	3.2.1.82	Poly-(1-4)-α-D-galactosiduronate digalacturonohydrolase.	Hydrolysis	Penultimate bonds.	Pectic acid	Digalacturonates.
5. Oligogalacturonate hydrolase			Hydrolysis	Terminal	Trigalacturonate	Monogalacturonates;
6. Δ4:5 Unsaturated oligogalacturonate hydrolases			Hydrolysis	Terminal	Δ4:5(Galacturonate)n	Unsaturated monogalacturonates & saturated (n-1).
7. Endopolymethyl-galacturonases			Hydrolysis	Random	Highly esterified pectin	Oligomethylgalacturonates
8. Endopolymethyl-galacturonases			Hydrolysis	Terminal	Highly esterified pectin	Oligogalacturonates.
b. Lyases						
1. Endopolygalacturonase lyase	4.2.2.2	Poly-(1-4)-α-D-galactosiduronate lyase.	Trans-elimination	Random	Pectic acid	Unsaturated oligogalacturonates.
2. Exopolygalacturonase lyase	4.2.2.9	Poly-(1-4)-α-D-galactosiduronate exolyase	Trans-elimination	Penultimate bond	Pectic acid	Unsaturated digalacturonates
3. Oligo-D-galactosiduronate lyase	4.2.2.6	Oligo-D-galactosiduronate lyase	Trans-elimination	Terminal	Unsaturated digalacturonates	Unsaturated monogalacturonates.
4. Endopolymethyl-D-galactosiduronate lyase	4.2.2.10	Poly(methyl galactosiduronate) lyase	Trans-elimination	Random	Unsaturated poly (methyl-D-digalacturonates).	Unsaturated methyloligogalacturonates.
5. Exopolymethyl-D-galactosiduronate lyase.			Trans-elimination	Terminal	Unsaturated poly-(methyl)-D-digalacturonates)	Unsaturated methylmonogalacturonates.

During the course of investigations and research conducted in recent years, the above classification of pectinolytic enzymes has slowly become obsolete. A more updated classification of these enzymes is presented in Table **1** [12].

Quantifying Pectinase Activity

The available commercial pectinase preparations used for juice processing generally contain a mixture of pectinesterase (PE), polygalacturonase (PG) and pectinlyase (PL) enzymes [13]. Complete pectin breakdown in juices can only be ensured if all three types of enzyme are present in the correct proportions.

Pectinase activity assays can be divided into three types: (1) the accumulation of products after hydrolysis; (2) the reduction in substrate quantity; and, (3) the change in the physical properties of substrates [12].

The majority of assays involve the accumulation of hydrolysis products, including reducing sugars, total sugars, and chromophores. The most common assay for reducing sugar determination is the dinitrosalicyclic acid (DNS) method described by Miller [14].

Depending on the type of substrate (*i.e.,* pectin, polygalacturonate or polymethylgalacturonate) and the enzyme mode of action (endo or exo activity), PGase activity can be quantified, and therefore expressed under different units, whether by the reduction of viscosity (endo-PG) in the reaction mixture or by the release of reducing groups (exo-PG) during the enzymatic reaction under established condition [15].

Pectinesterase (PE, Pectin pectlylhydrolase), often referred to as pectinmethylesterase, pectase, pectin methoxylase, pectin demethoxylase and pectolipase, is a carboxylic acid esterase and belongs to the hydrolase group of enzymes [16]. It catalyzes the deesterification of methyl ester linkages of galacturonan backbone of pectic substances to release acidic pectins and methanol.

Pectin lyases are the only known pectinases capable of degrading highly esterified pectins (like those found in fruits) into small molecules *via* β-elimination mechanism without producing methanol, in contrast with the combination of PG and PE, which are normally found in commercial products. This is important because methanol is toxic and may present health hazards [17].

Reducing Sugar Assay for Dinitrosalicylic Acid (DNS)

The reducing sugar assays that has been used in enzymes assays is the dinitrosalicylic acid (DNS) method [14], described below. This method tests for the presence of free carbonyl group (C=O), the so-called reducing sugars. This involves the oxidation of the aldehyde functional group present in, for example, glucose and the ketone functional group in fructose. Simultaneously, 3, 5-dinitrosalicylic acid (DNS) is reduced to 3-amino -5-nitrosalicylic acid under alkaline conditions (Fig. **3**).

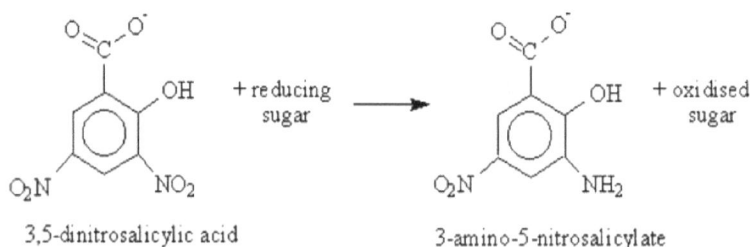

3,5-dinitrosalicylic acid   3-amino-5-nitrosalicylate

Figure 3: Reaction between 3,5-dinitrosalicylic acid (DNS) and reducing sugar. (Source: http://www.lsbu.ac.uk/biology/enzyme/practical1.htmL).

Protocol for Measuring Reducing Sugars by the DNS Method [14, 18, 19].

Materials

1. Distilled water;

2. 3,5 Dinitrosalicylic acid (DNS);

3. Sodium hydroxide;

4. Galacturonic acid;

5. Rochelle salts (sodium potassium tartrate);

6. Phenol;

7. Sodium metabisulfite;

8. Volumetric flasks -100, 1000 and 2000 mL;

9. Glass cuvette.

Equipment

1. Spectrophotometer for measuring absorbance at 540 nm;

2. Analytical balance;

3. Water bath capable of maintaining temperatures of 50 and100°C±0.1 °C.

Procedure for Preparing the DNS Reagent

1. Dissolve 10.6 g of DNS and 19.8 g of NaOH in 1,416 mL of distilled water;

2. After complete dissolution, add 360 g of Rochelle salts (sodium potassium tartrate), 7.6 mL of melted phenol (at 50°C), and 8.3 g of sodium metabisulfite, and then mix well;

3. Titrate 3 mL of the DNS reagent using 0.1 M HCl, phenolphthalein until the end point pH check. It should take 5–6 mL of HCl for a transition from red to colorless. Add NaOH if required (2.0 g of NaOH added corresponds to 1 mL of 0.1 M HCl used for 3 mL of the DNS reagent).

Procedure for Measuring Reducing Sugars

1. Place 1-2 mL sample in a test tube and add 3 mL DNS reagent;

2. Boil the content of all tubes for 5.0 min in a vigorously boiling water bath containing sufficient water to cover the portions of the tubes filled by the reaction mixture plus reagent. All samples, controls, blanks, and glucose standards should be boiled together;

3. After boiling, transfer to a cold ice water bath;

4. Dilute the content of all tubes (samples, blanks, standards and controls). Add 0.2 mL of color developed reaction mixture plus 2.5 mL of water in a spectrophotometer cuvette and mix well;

5. Measure absorbance at 540 nm;

6. Draw a standard sugar curve (sugar concentration along the x-axis *vs.* absorbance at 540 along the y-axis) following the volumes and standard concentrations of the enzyme assay. Different color intensities are generated by the different sugar concentrations, as shown in Fig. **4**.

Figure 4: Reducing sugar standard showing the color gradient generated by the different concentrations of sugar (Source: Cristiane Sanchez Farinas - *Embrapa Instrumentation*).

Modified DNS Reagent

The use of a modified DNS reagent has being adopted in some laboratories in order to avoid the use of phenol due to its toxicity. Since phenol is responsible to increase the amount of color produced during the color developing reaction, the values of absorbance obtained using this modified reagent is lower than the ones described before. Consequently, the slope of the curve is usually smaller.

Preparing the Modified DNS Reagent

1. Dissolve 300 g of sodium potassium tartarate tetrahydrate and 16 g of sodium hydroxide in 500 mL of distilled water;

2. When the solution is clear, add 10 g of DNS;

3. Bring the volume to 1000 mL with distilled water in a volumetric flask;

4. Store the solution in amber bottle (or dark bottle) protected from light;

5. Follow the same procedure described above for measuring reducing sugars.

Galacturonic Acid Standard Curve (Table 2)

1. Prepare a stock solution of galacturonic acid 10 µmol /mL (Dissolve 48mg (250 µmol) of galacturonic acid in 25 mL of distilled water;

2. Follow the procedure described in "Procedure for Measuring Reducing Sugars" obtain a standard curve;

3. The standard curve obtained must be plotted with galacturonic acid concentration along the x-axis *vs.* absorbance at 540 along the y-axis. From the standard curve slope and linear coefficients can be obtained, in order to determine the amount of reducing sugar released for each polygalacturonase enzyme tested.

Table 2: Galacturonic acid standard curve.

Tube (Number)	Solution of galacturonic acid (10µmol/mL) (mL)	Sodium Acetate Buffer pH 4.5 (mL)	Final concentration (µmol / mL)	DNS (mL)
1	0.0	1.0	0.0	1.0
2	0.2	0.8	2.0	1.0
3	0.4	0.6	4.0	1.0
4	0.6	0.4	6.0	1.0
5	0.8	0.2	8.0	1.0
6	1.0	0.0	10.0	1.0

Solutions

All solutions should be stored in glass flasks, labeled with date, concentration and responsible for preparation. Solutions must be kept under refrigeration and acclimatized before use. Validity of each solution must be accurately observed.

Sodium Acetate Buffer 0.2M

Materials

1. Acetic acid (CH_3COOH);

2. Sodium acetate trihydrate ($NaC_2H_3O_2\cdot 3H_2O$);

3. Distilled water;

4. Glass flask (250 mL); and,

5. Volumetric flasks -1000 mL, 100 mL.

Equipment

1. Analytical balance;

2. Potentiometer.

Procedures

How to prepare Sodium Acetate Buffer 0.2M:

Reagents must be mixed in different proportions, depending on the pH desired.

1. Solution (A), acetic acid 0.2M: Dilute 11.5 mL of acetic acid in 500mL of distilled water. In a 1000mL volumetric flask raise the volume with distilled water. Keep the solution in the refrigerator;

2. Solution (B), sodium acetate 0.2 M: Dissolve 16.4 g of $C_2H_3O_2Na$ or 27.2 g. $C_2H_3O_2Na. 3H_2O$ in 500 mL of distilled water, with the help of a magnetic stirrer. In a 1000mL volumetric flask raise the volume with distilled water. Keep the solution in the refrigerator;

3. For pH 4.5: mix 560 mL of solution (A) + (and) 440 mL of solution (B);

4. For pH 5.6: mix 96 mL of solution (A) + (and) 904 mL of solution (B).

Sodium Hydroxide Solution (NaOH 0,01M)

Materials

1. Sodium hydroxide (NaOH);

2. Distilled water; and,

3. Volumetric flasks -1000 mL.

Equipment

1. Analytical balance;

2. Potentiometer.

Procedures

1. Dissolve 0.4 g NaOH in 100 mL of distilled water;

2. After the complete dissolution, raise the volume to 1000 mL with distilled water in a volumetric flask; and transfer to glass container.

Polygalacturonases (Pgases) Assay

As previously already mentioned, Polygalacturonases (PGases) are the pectinolytic enzymes that catalyze the hydrolytic cleavage of the polygalacturonic acid chain with the introduction of water across the oxygen bridge. They are the most extensively studied among the family of pectinolytic enzymes. The PGases involved in the hydrolysis of pectic substances are endo-PGase (E.C. 3.2.1.15) and exo-PGase (E.C. 3.2.1.67).

PGase activity is determined on the basis of measuring, during the course of the reaction: (a) the rate of increase in number of reducing groups; and (b) the decrease in viscosity of the substrate solution [3].

Polygalacturonase Activity

The amount of reducing sugars can be readily measured by colorimetric methods like 3,5-dinitrosalicylate reagent method [14].

One unit of enzyme activity is defined as the amount of the enzyme that releases 1 µmol of galacturonic acid per min under standard assay conditions.

Materials

1. Polygalacturonic acid ($C_6H_{10}O_7$ $7H_2O$);

2. Enzyme diluted in acetate buffer solution pH = 4.5;

3. DNS reagent (Sodium hydroxide, potassium sodium tartrate);

4. Micro pipettes 0.1, 1.0 and 5.0 mL;

5. Timer;

6. Spectrophotometer tubes;

7. Test tubes; and,

8. Volumetric flasks - 100mL.

Equipment

1. Thermostatic bath at 35° C ±-0.1 °C;

2. Water bath at 100°C ±0.1 °C;

3. Spectrophotometer at 540 nm;

4. Analytical balance; and

5. Potentiometer.

Polygalacturonic Acid Solution 0.25% in Acetate Buffer, pH 4.5.

Weigh 0.25 g polygalacturonic acid in analytical balance and dissolve in 50 mL of acetate buffer solution at pH 4.5; after the complete dissolution, raise the volume to 100mL in a volumetric flask; transfer to a glass container.

Protocol for Enzyme Assays

1. In 4 mL of 0.25% polygalacturonic acids solution add 0.1 mL of enzyme solution (diluted if needed), and then incubate at 35° C for 30 minutes;

2. At the end of the incubation period, pipette 0.5 mL of the reaction mixture in an assay tube and stopping the enzyme reaction by immediately adding 0.5 mL of DNS reagent;

3. Boil the content of all tubes for exactly 5.0 minutes in a vigorously boiling water bath containing sufficient water to cover the portions of the tubes filled by the reaction mixture plus reagent;

4. All samples, controls and blanks, should be boiled together;

5. After boiling, transfer the tubes to a cold ice water bath; add 5.5mL of distilled water;

    Measure absorbance at 540 nm for all samples. Absorbance of all samples must fall in the range from 0.1 to 0.7 A.

Blank and Enzyme Controls

1. Enzyme control test: add 0.1 mL enzyme inactivated of inactivated enzyme (diluted in acetate buffer solution at pH 4.5 if necessary, and inactivated by boiling in water for 20 minutes) in tubes containing 4 mL of 0.25% polygalacturonic acid solution, incubate at 35° C, for 30 minutes;

2. Blank test: add 0.5 mL of distilled water and 0.5 mL of DNS reagent.

Determination of Enzyme Activity

As previously said, one unit of enzyme activity is defined as the amount of the enzyme that releases 1 μmol of galacturonic acid for min under standard assay conditions.

1. With the help of the standard curve obtained, it is possible to calculate the true galacturonic acid concentrations, after release by each sample;

2. The enzyme concentration is expressed in UI (μmol of product released per minute)/mL.

Where:

UI = dilution factor x total volume (mL) x galacturonic acid (µmol/mL)/ reaction time (min) x volume enzyme (mL)

Galacturonic acid concentration (x) (µmol/mL) of unknown sample is:

$$(x) = y \pm \frac{b}{a}$$

Where: y = absorbance found

a= angular coefficient of straight

b = linear coefficient of straight

Calculation

Using this standard curve, convert the absorbance values of the sample tubes (after subtraction of enzyme blank) into galacturonic acid (= mg of galacturonic acid produced during the reaction).

Polygalacturonase Activity Measured by the Decrease in Viscosity of the Substrate Solution

The PGase activity can be determined also by viscosity reduction of pectin solution using a viscosimeter [20]. However, this method still has a limited application. There is no direct correlation between viscosity reduction and the number of glycosidic bonds hydrolyzed. PGase activity can also be determined by the cup-plate method [21].

In that case the unit used for enzyme activity is the amount of enzyme required to obtain a certain degree of viscosity per unit of time.

Materials

1.   Citric pectin;

2.   Acetate buffer solution at pH 4.5;

3.   Enzyme diluted in acetate buffer solution at pH 4.5;

4.   Timer;

5.   Thermometer;

6.   Micro pipettes 0.1, 1.0 and 5.0 mL;

7.   Graduated cylinder of 50 mL; and,

8.   Volumetric flasks - 100mL.

Equipment

1.   Refrigerated thermostatic bath;

2.   Thermostatic bath at 35° C ±-0.1 °C;

3.   Spectrophotometer at 540 nm;

4.   Analytical balance;

5.   Potentiometer; and,

6.   Viscosimeter.

Citric Pectin Solution 0.63% in Acetate Buffer Solution at pH 4.5

To obtain prepare 100 mL of solution weight, in analytical balance, 0.63 g citric pectin and dissolve in 50 mL of acetate buffer solution, at pH 4.5; after the complete dissolution, raise the volume to 100 mL in a volumetric flask; transfer the contents to glass container.

Enzyme Assays

1.   The reaction mixture consisted of 14.8 mL of citric pectin solution (0.63% w/v) and 3.2 mL of appropriately diluted enzyme source;

2.   After incubation at 30°C for 10 min, the viscosity of the reaction is measures using a viscosimeter [22].

Viscosimeter Conditions

The viscosity data can be obtained by Brookfield Viscosimeter, model RVDV III ULTRA (Fig. **5**), with concentric cylinders and using a SC4-21 spindle, keeping the temperature at 30°C and the rate of shearing stress between 46.5 and 134.5 s^{-1}.

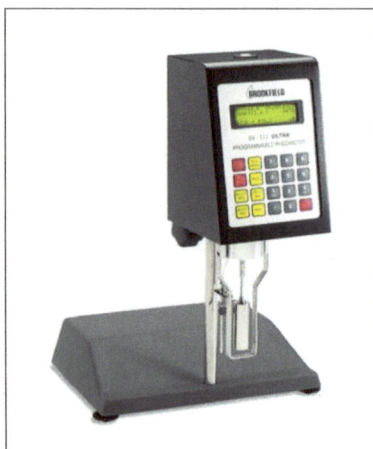

Figure 5: Brookfield Viscosimeter, model RVDV III ULTRA (Source: http://www.labsource.co.uk/shop/brookfield-rvdviii-ultra-programmable-rheometers-p-1402.html, April 2011).

Reduction of the Viscosity Rate

The rate of viscosity reduction (RVR) is calculated using the equation below, where T, Ta and T0 represent the flow rate (in a capillary viscometer) in seconds (s) for the reaction mixture without enzyme (T), the test mixture (Ta) and water (T0), respectively.

$$\text{RVR (\%)} = \frac{T_a(s) - T(s)}{T_a(s) - T_0(s)} * 100$$

The unit of enzyme activity is mostly selected as the amount of enzyme required for attaining a certain decrease of viscosity per unit time.

UI = enzyme activity

UI = Viscosity reduction (RVR) / reaction time (min.) x volume enzyme (mL)

Due to Newtonian fluid behavior of pectin solution, the viscosity can be calculated as the ratio (constant of proportionality) between shear stress (Poise) and rate of shearing stress (s^{-1}).

Pectinlyase (pl) Activity Assay

Lyases (or transeliminases) perform non-hydrolytic breakdown of pectates or pectinates, characterized by a trans-eliminative split of the pectic polymer [23]. The lyases break the glycosidic linkages at C-4 and simultaneously eliminate H from C-5, producing a Δ4:5 non-saturated product of the substrate solution (Fig. **6**) [24].

The activity of pectin lyase is assayed by measuring the increase in optical density at 235 nm due to the formation of 4,5-unsaturated oligogalactouranotes by β-elimination mechanism molar extinction coefficient for the product being 5.5×10^3 $M^1 cm^1$ at λ= 235 nm [24].

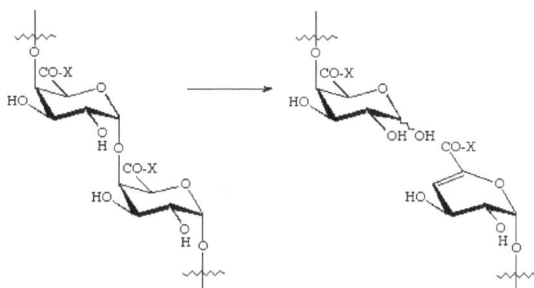

Figure 6: Mode of action of Pectilyase, for pectate X is predominantly -O- and for pectin X is predominantly -O-Me (Source: http://www.enzyme-database.org/reaction/polysacc/4222.htmL, April 2011).

One unit of enzymatic activity (U) is defined as the amount of enzyme which released 1 μmol of unsaturated uronide per minute, based on the molar extinction coefficient (5500) of the unsaturated products. The enzyme production was expressed in units per mL (U/mL).

Operating Procedure – Determination of Pectinlyase (PL) Activity

Materials

1. Apple pectin;

2. Calcium chloride;

3. Solution 0.5M H_2SO_4;

4. Acetate buffer solution pH = 5.5;

5. Enzyme diluted in acetate buffer solution pH = 5.5;

6. Micro pipettes 0.1, 1.0 and 5.0 mL;

7. Volumetric flask -100 mL;

8. Test tubes;

9. Chronometer;

10. Thermometer; and,

11. 1 cm lightpath quartz cuvette.

Equipment

1. Thermostatic bath at 35° C ±-0.1 °C, with stirring;

2. Spectrophotometer at 235 nm.

Apple Pectin Solution 1.0% in Acetate Buffer Solution at pH 5.5

To obtain 100 mL of solution: weight in analytical balance 0.5 g of apple pectin and dissolve in 50 mL of acetate buffer solution, at pH 5.5; after the complete dissolution, raise the volume to 100 mL in a volumetric flask; transfer to glass container.

Enzymatic Assay

1. The reaction mixture consisted of 2.5 mL of 0.5% apple pectin solution, 0.5 mL calcium chloride, 21.0 mL of sodium acetate buffer solution pH 5.5, acclimated at 35° C for 10 minutes and 1.0 mL of enzymatic extract (diluted in acetate buffer solution at pH 5.5); incubates at 35° C, for 15 minutes;

2. When the crude enzyme solution added to pectin solution the enzyme starts to break the glycosidic bonds of pectin by β-elimination;

3. Due to this action the solution becomes turbid. The increase in A_{235nm} is recorded for 15 minutes;

4. Blank tests are must be performed by adding the acid previously to the enzyme (no reaction expected to happen); and,

5. Distilled water to reset the spectrophotometer and the samples are read against blank.

Then:

$$UI = (A235test - A235 \text{ blank}) (25) (d_f)/ \text{ reaction time (min) x volume enzyme (mL)}$$

$$25 = \text{Total volume (mL) of assay}$$

$$d_f = \text{dilution factor}$$

$$15 = \text{reaction time (min)}$$

$$= \text{Volume (mL) of enzyme used}$$

Pectinesterase (Pe) Activity Assay

Pectinesterase (PE, Pectin pectlylhydrolase, E.C. 3.1.1.11), often referred to as pectinmethylesterase, pectase, pectin methoxylase, pectin demethoxylase and pectolipase, is a carboxylic acid esterase and belongs to the hydrolase group of enzymes [25]. It catalyzes the deesterification of methyl ester linkages of galacturonan backbone of pectic substances to release acidic pectins and methanol [26].

PE activity can be measured by using a pH stat because ionization of the carboxyl group of the product releases a proton, which causes a change in pH [25, 27].

One Unit of PE activity was taken as the amount of NaOH (mEq) consumed per min to keep constant pH value.

Operating Procedure for Determination of Pectinesterase (PE) Activity-Titrimetric Assay

Materials

1. Citrus fruit pectin (degree of esterification 68%);

2. Sodium Hydroxide (NaOH);

3. Enzyme extract;

4. Test tubes;

5. Chronometer;

6. Thermometer.

Equipment

1. Potentiometer;

2. Thermostatic bath at 35° C +-0.1 °C, with stirring;

3. Thermostatic bath at 100° C +-0.1 °C, with stirring.

Apple Pectin Solution 1.0%

To obtain 100 mL of solution weight, in analytical balance, 0.5 g apple pectin and dissolve in 50 mL of water, after the complete dissolution, raise the volume to 100 mL in a volumetric flask. The pH is adjusted to 6.5 with 0.01M NaOH.

Enzyme Assay

PE activity is measured by determining the carboxyl groups released by titration with 0.01M NaOH (Jiang *et al*, 2001 with modifications).

1. One mL of enzyme solution is added to 15 mL of 0.01M NaCl/0.5% pectin citrus solution, with the pH brought adjusted to 6.5 immediately before assay;

2. The free protons dissociated from the free carboxyl groups formed is measured by titrating using an autotitrator;

3. The volumes (mL) of 0.01 N NaOH consumed to maintain a pH of 6.0 of the reaction solution at 35^0C (in a water bath) are recorded within 10 min;

4. Enzyme control test: One mL of enzyme inactivated by boiling in water for 20 minutes, is added to 15 mL of 0.1M NaCl/0.5% citrus pectin solution, with the pH brought adjusted to 6.5 immediately before assay;

5. The volumes (mL) of 0.01 N NaOH consumed to maintain a pH of 6.0 of the reaction solution at 35^0C (in a water bath) are recorded within 10 min.

Then:

$$UI = (Vol\ NaOH\ test - Vol\ NaOH\ blank)\ (dilution\ factor)/\ reaction\ time\ (min)\ x\ volume\ enzyme\ (mL)$$

CONCLUSIONS

Several assay methods based on enzymatic reactions using viscometric, colourimetric, spectrophotometric, or pH-titration detection of the reaction products presented in this chapter have been tested by different laboratories and represent the most accurate and sensitive methods for the determination of activity of pectinolytic enzymes in complex mixtures of enzymes.

ACKNOWLEDGEMENTS

The authors thank Embrapa and CNPq for the financial support to conduct this work.

CONFLICT OF INTEREST

The authors confirm that this chapter content has no conflict of interest.

REFERENCES

[1]   PILNIK, W.; VORAGEN, A. G. J. Pectic Substances and other Uronides. In: The Biochemistry of Fruits and their Products. Hulme, A.C. (Ed.). Academic Press, London, 1970, pp: 53-87.
[2]   JAYANI, R. S.; SAXENA, S.; GUPTA, R. Microbial pectinolytic enzymes: A review. Process Biochemistry 2005, 40:2931-2944
[3]   REXOVA BENKOVA, L.; MARKOVIC, O. Pectic enzymes. Advances in Carbohydrate Chemistry, 1976, 33:323–85.
[4]   ALKORTA, I. C.; GARBISU, M. J.; LIAMA AND J.L. Serra. Industrial applications of pectic enzymes: A review. Process Biochemistry, 1998, 33:21-28.

[5]   KASHYAP, D.R.; VOHRA, P.K.; CHOPRA, S.; TEWARI, R.1. Source: Applications of pectinases in the commercial sector: a review. Bioresource Technology, 2001, 77:215-227.

[6]   PARIZA, M. W.; JOHNSON, E. A., Evaluating the safety of microbial enzyme preparations used in food processing: update for a new century. Regulatory Toxicology and Pharmacology, 2001, 33:173±186.

[7]   NAIDU, G. S. N.; PANDA, T., 1998, Production of pectolytic enzymes - A review. Bioprocess Engineering, 1998, 189:355–61.

[8]   SATHYANARAYANA, N.; GUMMADI, T. Panda. Purification and biochemical properties of microbial pectinases-a review. Process Biochemistry, 2003, 38:987-996.

[9]   SAKAI, T. Degradation of pectins. In: Winkelmann G, editor. Microbial degradation of natural products. Weinheim: VCH; 1992. p. 57–81.

[10]   PALOMAKI, T.; SAARILAHTI, H. T. Isolation and characterization of new C-terminal substitution mutation affecting secretion of polygalacturonases in Erwinia carotovora ssp. carotovora. FEBS Letters, 1997, 400:122–126.

[11]   ALBERSHEIM, P.; NEUKOM. H.; DEUEL, H. Splitting of pectin chain molecules in neutral solution. Archives of Biochemistry and Biophysics, 1960, 90:46–51.

[12]   CODNER, R.C. Pectinolytic and cellulolytic enzymes in the microbial modification of plant tissues. Journal of Applied Bacteriology, 2001, 84:147-160.

[13]   RANVEER SING, SAXENA S., GUPTA R. Microbial pectinolytic enzyme: A review. Process Biochemistry, 2005, 40: 2931-2944.

[14]   MILLER, G. L. Use of dinitrosalicylic acid reagent for determination of reducing sugars. Analytical Chemistry, 1959, 31:426–428.

[15]   FAVELA-TORRES, E.; VOLKE-SEPÚLVEDA, T.; VINIEGRA-GONZÁLEZ. G. Hydrolytic Depolymerising Pectinases, Food Technology and Biotechnology, 2006, 44:221–227.

[16]   WHITAKER, J. R. Pectic substances, pectic enzymes and haze formation in fruit juices. Enzyme and Microbial Technology 1984, 6:341–347.

[17]   YADAV, S.; YADAV, P. K.; YADAV, D.; KAPIL, D. E. O.; SINGH YADAV, K. D. S. Pectin lyase: A review. Process Biochemistry, 2009, 44:1–10.

[18]   GHOSE, T. K. Measurement of cellulase activities. Pure and Applied Chemistry 1987, 59:257-268.

[19]   ZHANG Y-HP.; LYND, L. R. Kinetics and relative importance of phosphorolytic and hydrolytic cleavage of cellodextrins and cellobiose in cell extracts of *Clostridium thermocellum*. Applied and Environmental Microbiology 2004, 70:1563–1569.

[20]   ROBOZ, E.; BARRATT, R. W.; TATUM, E.L. Breakdown of pectic substances by a new enzyme from Neurospora. Journal of Biological Chemistry, 1952, 195:459-471.

[21]   DINGLE, J.; W. W. REID; SOLOMONS, G. L. The enzymatic degradation of pectin and other polysaccharides. II. Application of the 'Cup-plate' assay to the estimation of enzymes. Journal of Science of Food and Agriculture, 1953, 4:149-155.

[22]   FONTANA, R.C.; POLIDORO, T.A.; SILVEIRA, M.M. Comparison of stirred tank and airlift bioreactors in the production of polygalacturonases by *Aspergillus oryzae*. Bioresource Technology, 2009, 100: 4493-4498.

[23]   SAKAI, T.; SAKAMOTO, T.; HALLAERT, J.; VANDAMME, E. J. Pectin, pectinase and protopectinase: production, properties and applications. Advances in Applied Microbiology, 1993, 39:231–94.

[24]   ALBERSHEIM, P. Pectin lyase from fungi. Methods in Enzymology, 1966, 8:628–31.

[25]   COSGROVE, D.J., Assembly and enlargement of the primary cell wall in plants. Annual Review of Cell Biology, 1997, 13:171-201.

Send Orders for Reprints to reprints@benthamscince.net

CHAPTER 2

Assays of Peroxidase Activity

Bernardo Dias Ribeiro[1],* and Maria Alice Zarur Coelho[2]

[1]School of Chemistry, Federal University of Rio de Janeiro, Rio de Janeiro, Brazil and [2]Departament of Biochemical Engineering, Biological Systems Engineering, School of Chemistry, Federal University of Rio de Janeiro, Rio de Janeiro, Brazil

Abstract: Peroxidases are enzymes that use various peroxides as electron acceptors in oxidation, and are widespread in nature. Their activities are usually determined by spectrophotometric methods, which a condensation or oxidation product is formed in the presence of H_2O_2, or by titrimetric methods, using potassium permanganate or sodium thiosulfate, which measure free hydrogen peroxide in solution.

Keywords: Peroxidase, hydrogen peroxide, alkyl peroxides, benzyl peroxide, catalase, lactoperoxidase, myeloperoxidase, horseradish peroxidase, vanadium bromoperoxidase, lignin peroxidase, glutathione peroxidase, cytochrome C peroxidase, guaiacol.

INTRODUCTION

Peroxidases are enzymes that use various peroxides (ROOH), such as hydrogen peroxide, alkyl peroxides, and benzyl peroxide, as electron acceptors to catalyze a number of oxidative reactions, besides epoxidation and enantioselective reduction of racemic hydroperoxides. They are extremely widespread with more than 7000 enzymes listed and present in plants, microorganisms, and higher organisms. In mammals, they are implicated in biological processes as immune system or hormone regulation. In plants, they are involved in auxin metabolism, lignin and suberin formation, cross-linking of cell wall components, defense against pathogens or cell elongation. Humans contain more than 30 peroxidases whereas *Arabidopsis thaliana* has about 130 peroxidases [1,2].

These peroxidases can be heme and non-heme proteins, and can be found in the

---

*Address correspondence to Bernardo Dias Ribeiro: School of Chemistry at Federal University of Rio de Janeiro, Brazil; Tel: 55 (21) 2562-7646; Fax: 55 (21) 2562-7567; E-mail: dias.bernardo@gmail.com.

sub-subclass E.C.1.11.1.x, donor: hydrogen- peroxide oxidoreductase. Most heme peroxidases belong to two large families, one mainly found in plants and also in bacteria and fungi, and a second one which is found mostly in animals; and also as four smaller protein families: catalases, di-heme cytochrome C peroxidases, dyp-type peroxidases and heme haloperoxidases. Non-heme peroxidases are not evolutionarily linked and form five independent families: thiol peroxidase (the largest one, which currently contains more than 1000 members), alkylhydroperoxidase, non-heme haloperoxidase, manganese catalase and NADH peroxidase [2]. In Table **1**, the properties of some peroxidases are shown.

Table 1: Properties of major peroxidases.

Peroxidases	Active Center	Sources	Molar Mass (kDa)	Function
Catalase [3]	Heme-group	Animals, Plants, microorganisms	60 - 75	Primary component of the antioxidant system, acting against oxidative stress
Lactoperoxidase [4]	Heme-group	Milk, Saliva, Tears	78	One of the nonimmunoglobulin defense factors in the mucosal secretions, and antimicrobial
Myeloperoxidase [5]	Heme-group	Neutrophils and macrophages	84 - 89	Biomarker in acute coronary syndromes, and antimicrobial
Horseradish peroxidase [6]	Heme-group	*Armoracia rusticana*	36 - 38	Biosynthesis of plant hormones, lignification, crosslinking of cell wall polymers, suberin formation and resistance to infection
Vanadium Bromoperoxidase [7]	Non-heme group (Vanadium)	Marine algae	60 - 75	Synthesis of halogenated natural products with antimicrobial activity
Lignin Peroxidase [8]	Heme-group	*Phanerochaete chrysosporium* and other white-rot fungi	38 - 46	Degradation and oxidation of lignin and phenolic substrates
Glutathione Peroxidase [9]	Non-heme group (Selenium)	Animals, Plants, microorganisms	18 - 22	Protects cells against membrane lipid peroxidation and cell death.
Cytochrome C Peroxidase [10]	Heme-group	Mitochondria	34 - 37	Protects the organism from high concentrations of hydrogen peroxide in aerobic respiration

Most peroxidases, more than 70%, are heme enzymes and contain the ferric protoporphyrin IX (protoheme) group with an iron atom in their active center. The group is held in position by electrostatic interaction of one propionic acid group of the heme and a lysine residue (Lys174) of the apoprotein. Two structural Ca^{2+} ions are commonly observed in the folded molecule, and are considered necessary for its stability and also as an activator of plant peroxidases [11,1] (Barceló & Pomar, 2002; Riehmann & Ritter, 2006).

The catalytic cycle of heme peroxidases, as horseradish peroxidase, may be described in Fig. **1**. The first step is the formation of a precursor complex of hydrogen peroxide (or an organic hydroperoxide) with the enzyme (**1**).

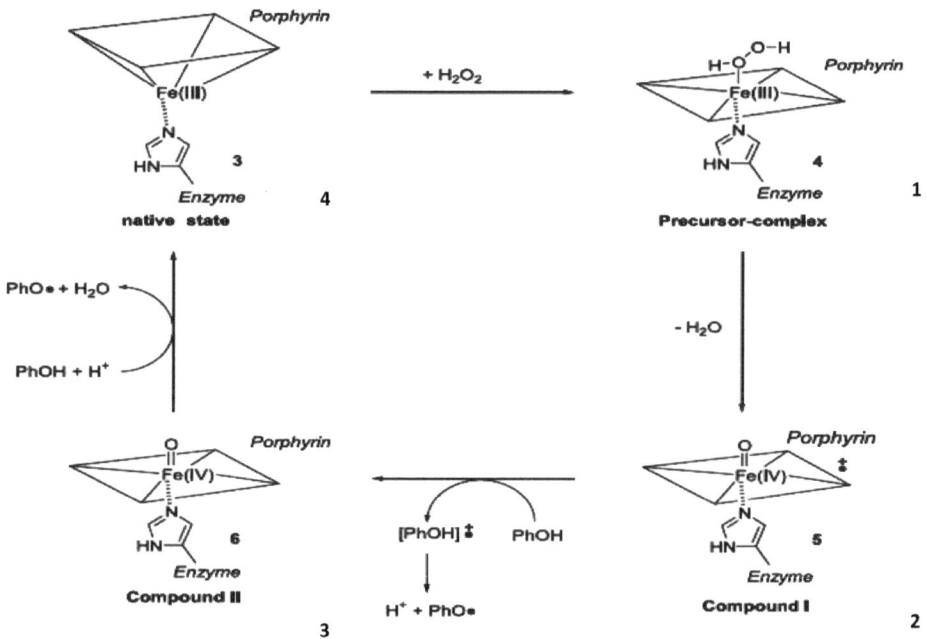

Figure 1: Catalytic cycle of horseradish peroxidase [1].

Next, the elimination of water leads to the so-called compound I (**2**), with an oxyferryl group in the active center surrounded by porphyrin in the form of a Π radical cation. Fe (III) of the active center of HRP is oxidized to Fe (IV) while the porphyrin structure loses one electron. Compound I oxidizes the first equivalent of phenol by oxidoreduction. The phenol coordinates in the active center and is oxidized *via* one-electron transfer to the corresponding phenoxy radical cation, while the porphyrin radical cation is reduced to the neutral porphyrin. After deprotonation of the phenoxy radical cation, the resulting phenoxy radical leaves the active center, with an irreversible ping-pong mechanism. The proton of the phenoxy radical is abstracted by a base, probably distal histidine (His – 42) to form the so-called compound II (**3**).

This species oxidizes the second equivalent phenol. The phenol is oxidized to the phenoxy radical by one-electron transfer to the Fe (IV), so that the native enzyme is regenerated (**4**). The proton of the phenol forms water as a leaving group with the oxygen of the oxyferryl group and the proton which was bound to the His – 42. Thus, in one catalytic cycle, two equivalents of a phenol are oxidized by one equivalent of hydrogen peroxide producing two phenoxy radicals and one water molecule [1, 11].

Based on the mechanism above presented some Peroxidase Activity Measurements will be presented:

Peroxidase Assay

Principle of the Method

The enzymatic peroxidase activity is usually determined by a colorimetric method, based on the change of absorbance at 430 - 510 nm due to: formation of oxidation product from an analyte as guaiacol [12,13] o-dianisidine [14] or pyrogallol [15] condensation product as in the reactions of phenol-4-sulphonic acid and 4-aminoantipyrine [16] or pyrocatechol and aniline [17]in the presence of H_2O_2.

One of most used analyte is guaiacol (o-methoxyphenol), which generates amber colored product described as tetraguaiacol, 2,2'-dihydroxy-3,3'-dimethoxybiphenyl, 4,4'-dihydroxy-3,3'-dimethoxybiphenyl,   or   3,3'-dimethoxy-4,4'-biphenoquinone [12], as showed at Fig. **2**.

Procedure

Reagents

1. Sodium dihydrogen phosphate monohydrate;

2. Sodium hydrogen phosphate heptahydrate;

3. Hydrogen peroxide 30% v/v;

4. Guaiacol;

5. Sodium hydroxide (pearls);

6. Phosphoric acid;

7. Enzyme extract;

8. Distilled water.

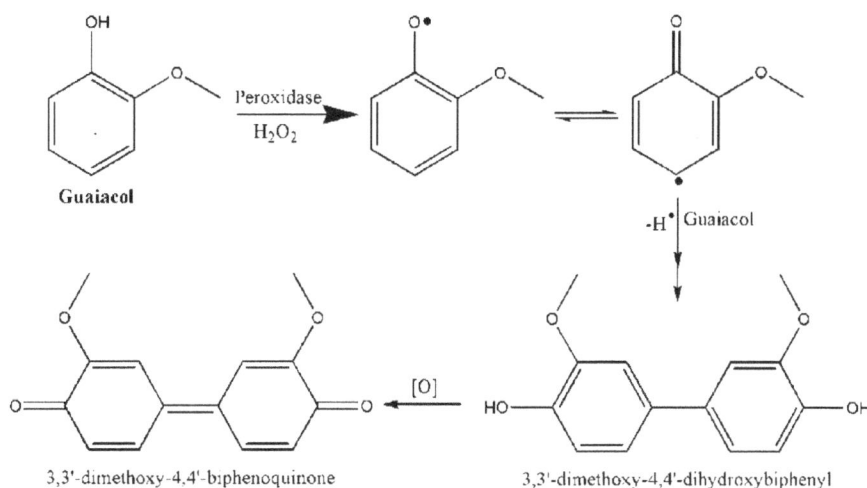

Figure 2: Oxidation of guaiacol by peroxidase.

Materials

1. Plastic microtubes (2mL, polypropylene);

2. Pipettes (10 µL, 100 µL, 1 mL and 5 mL);

3. Glass flask (100 mL);

4. Glass cuvette;

Equipment

1. Spectrophotometer (visible range) coupled to a computer;

2. Thermostatic bath (preferably digital, 0.1 °C precision);

3. Vortex mixer

Preparation of Solutions

Phosphate buffer 100 mM pH 6.0

Dissolve 1.2143 g of sodium dihydrogen phosphate monohydrate and 0.3218 g of sodium hydrogen phosphate heptahydrate in 100 mL distilled water to obtain 100 mL of the buffer at the pH 6.0. Check in a pH meter and if the pH deviates from the desired value, slowly add 7M phosphoric acid or 7M sodium hydroxide solution to reach pH 6.0.

Guaiacol Solution 100 mM

For the preparation of guaiacol solution, 1.2414 g (1.116 mL) of guaiacol was added to 100 mL of phosphate buffer 100 mM pH 6.0. The solution should be stored refrigerated for no longer than 2 weeks in amber glass flasks, labeled with date of preparation as well as solution information and operator identifications.

Hydrogen Peroxide Solution 2.0 mM

For the preparation of hydrogen peroxide solution, 15.6 μL of 30% hydrogen peroxide was cautiously added to 90 mL of phosphate buffer 100 mM pH 6.0, and diluted to 100 mL. The solution should be stored refrigerated for no longer than 1 week in amber glass flasks, labeled with date of preparation as well as solution information and operator identifications.

Operation

Assay of Enzyme Extracts [18]

For the determination of peroxidase activity in enzyme extracts with unknown activities, add 2.8 mL of phosphate buffer 100 mM pH 6.0, 100 μL of the guaiacol solution 100 mM, 100 μL of 2.0mM hydrogen peroxide solution and 20 μL of enzyme preparation to a glass cuvette, at 25 °C. Immediately, the reaction progress must be followed in a spectrophotometer, at 470 nm, coupled to a computer, for 100 s. Control experiments (blank runs) were always carried out using the same procedure, but in the absence of peroxidase.

Calculation of Peroxidase Activity

For the calculation of activity, use Equation 1. The range of absorbance to be used should be corresponding to the initial velocity of enzymatic reaction, thus, where there is a linear relationship of product formation with time. One unit of peroxidase activity (POD) is defined as representing the amount of enzyme enough to produce 1 μmol of product per minute per milliliter, under the assay conditions. This increase is proportional to the rate of H_2O_2 consumption.

$$POD\ (U\ mL^{-1}) = \frac{\Delta Abs}{\varepsilon \times l \times \Delta t \times V_{ENZ}} \tag{1}$$

Where:

ε is the molar absorption coefficient (mM-1.cm-1);

l is the path length (1 cm);

Abs is the absorbance values registered for spectrophotometer;

t represents the time of enzymatic reaction (min); and

VENZ represents the volume of enzyme extract (mL) used for the assay.

For ε determination of 3,3'-dimethoxy-4,4'-biphenoquinone produced from the unknown peroxidase reaction, it is necessary to make a graphic correlating the slope of linear plots of the maximum absorbance increase *versus* the number of hydrogen peroxide molecules consumed in enzymatic reactions, being hydrogen peroxide the limiting reagent (25 - 200 μM) and assuming a total and equimolar reaction. The hydrogen peroxide concentrations can be determined by monitoring their absorbance at 240 nm and using $\varepsilon = 43.6\ M^{-1}.cm^{-1}$.

For example, using the horseradish peroxidase, the ε is 5.53 mM^{-1}.cm^{-1}, and with myeloperoxidase, ε = 5.58 mM^{-1}.cm^{-1} [19]

Catalase Assay

Principle of the Method

The catalase activity can be determined by spectrophotometric methods, based on the decrease of absorbance at 240 nm due to H_2O_2 consume[20] or on the inhibition of color development in the assay using uricase and peroxidase which is H_2O_2 dependent [3]; or by titrimetric methods, using potassium permanganate or sodium thiosulfate [21].

The iodometric titration with sodium thiosulfate measure free hydrogen peroxide in solution, meaning the hydrogen peroxide which was not cleaved during enzymatic reaction. This hydrogen peroxide reacts with iodide in acid medium, then iodine liberated complexes with starch and forming a blue color. The thiosulfate solution will react with iodine and the color will disappear, as in Fig. **3**.

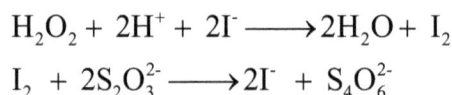

$$H_2O_2 + 2H^+ + 2I^- \longrightarrow 2H_2O + I_2$$
$$I_2 + 2S_2O_3^{2-} \longrightarrow 2I^- + S_4O_6^{2-}$$

Figure 3: Iodometric titration of hydrogen peroxide.

Procedure

Reagents

1. Sodium dihydrogen phosphate monohydrate;

2. Sodium hydrogen phosphate heptahydrate;

3. Hydrogen peroxide 30% v/v;

4. Sodium thiosulfate pentahydrate;

5. Sodium hydroxide (pearls);

6. Phosphoric acid;

7.   Potassium Iodide;

8.   Soluble starch;

9.   Ammonium molybdate;

10. Enzyme extract;

11. Distilled water;

12. Milli-Q water.

Materials

1.   Burette (50 mL);

2.   Plastic microtubes (2mL, polypropylene);

3.   Pipettes (10 μL, 100 μL, 1 mL and 5 mL);

4.   Glass flask (50, 100 and 200 mL);

5.   Thermostatic bath (preferably digital, 0.1 °C precision);

6.   Magnetic stirrer.

Preparation of Solutions

Phosphate Buffer 100 mM pH 7.0

Dissolve 0.5836 g of sodium dihydrogen phosphate monohydrate and 1.5466 g of sodium hydrogen phosphate heptahydrate in 100 mL distilled water to get 100 mL of the buffer at the pH 7.0. Check in a pH meter and if the pH deviates from the desired value, slowly add 7M phosphoric acid or 7M sodium hydroxide solution to reach pH 7.0.

Hydrogen Peroxide Solution 5% v/v

For hydrogen peroxide solution preparation, 5 mL of 30% hydrogen peroxide was cautiously added to 70 mL of phosphate buffer 100 mM pH 7.0, and diluted to 100 mL. The solution should be stored refrigerated for no longer than 1 week in amber glass flasks, labeled with date of preparation as well as solution and operator identifications.

Sodium Thiosulfate Solution 0.25 M

Dissolve 62.5 g of sodium thiosulfate pentahydrate in 750 mL Milli-Q water, add 3.0 mL of sodium hydroxide 0.2 M as a stabilizer, and diluted to 1000 mL with water.

Starch Solution 1% w/v

Add 1.0 g of soluble starch in 5.0 mL water, and stir until a paste is formed. Then add to paste 100 mL of boiling water. Cool and add 5.0 g of potassium iodide. Stir until dissolution is complete and transfer to a plastic bottle.

Operation

Assay of Enzyme Extracts [22]

For the determination of catalase activity in enzyme extracts with unknown activities, add 1.0 mL of enzyme preparation, previously diluted with phosphate buffer 100 mM pH 7.0, into a 200 mL becher. Immediately add 100 mL of hydrogen peroxide solution 5% v/v, at 25°C, and stir for 1 hour. Then pipet 4.0 mL from the becher into a 50 mL Erlenmeyer flask, and add 5.0 mL sulfuric acid 2M, 5.0 mL potassium iodide 40% w/v (freshly prepared) and 20 μL of ammonium molybdate 1% w/w, always stirring and tritrating with sodium thiosulfate solution 0.25 M until yellow pale color. In this moment, add 50 μL of starch solution to formation an intense blue color. Titrate with sodium thiosulfate solution until colourless endpoint, recording the required volume, in mL (S). Perform a blank determination changing the sample for 4.0 mL of hydrogen peroxide solution 5% v/v, and record the required volume, in mL, as B, which is usually 16 mL. If the blank volume was less than 14 mL, the hydrogen peroxide solution should be prepared fresh again.

Calculation of Peroxidase Activity

For activity calculation, use Equation 2. One unit of catalase activity (Baker Unity) is defined as representing the amount of enzyme enough to decompose 266 mg of hydrogen peroxide, under the assay conditions.

$$BU.mL^{-1} = 0.4 \times \frac{(B-S)}{V_{ENZ}} \tag{2}$$

Where:

B is the titration volume for blank (mL);
S, titration volume for sample (mL); and
V_{ENZ} represents the volume of enzyme extract (mL) used for the assay.

Equipment

Spectrophotometer.

ACKNOWLEDGEMENTS

The authors thank the financial support of the Brazilian National Council for Scientific and Technological Development (CNPq), (MCT- CNPq) and the Carlos Chagas Filho Foundation for Research Support in the State of Rio de Janeiro (FAPERJ).

CONFLICT OF INTEREST

The authors confirm that this chapter content has no conflict of interest.

REFERENCES

[1] RIEHMANN, M.; RITTER, H.; Synthesis of Phenol Polymers Using Peroxidases. Advances in Polymer Science, 2006; 194, 1-49.
[2] KOUA, D.; CERUTTI, L.; FALQUET, L.; SIGRIST, C.J.A.; THEILER, G.; HULO, N.; DUNAND, C.; PeroxiBase: a database with new tools for peroxidase family classification. Nucleic Acids Research, 2009; 37, D261-D266.
[3] SLAUGHTER, M.R.; O'BRIEN, P.J.; Fully-Automated Spectrophotometric Method for Measurement of Antioxidant Activity of Catalase. Clinical Biochemistry, 2000; 33 (7), 525-534.
[4] SEIFU, E.; BUYS, E.M.; DONKIN, E.F.; Significance of the lactoperoxidase system in the dairy industry and its potential applications: a review. Trends in Food Science & Technology, 2005; 16, 137–154,
[5] LORIA, V.; DATO, I.; GRAZIANI, F.; BIASUCCI, L.M.; Myeloperoxidase: A New Biomarker of Inflammation in Ischemic Heart Disease and Acute Coronary Syndromes. Mediators of Inflammation, 4 pages, Article ID 135625, 2008.

[6]   VEITCH, N.C.; Horseradish peroxidase: a modern view of a classic enzyme. Phytochemistry, 2004; 65, 249–259.

[7]   BUTLER, A.; CARTER-FRANKLIN, J.N.; The role of vanadium bromoperoxidase in the biosynthesis of halogenated marine natural products. Natural Products Reports, 2004; 21, 180-188.

[8]   WONG, D.W.S.; Structure and Action Mechanism of Ligninolytic Enzymes. Applied Biochemistry and Biotechnology, 2009; 157, 174-209.

[9]   BATTIN, E.E.; BRUMAGHIN, J.L.; Antioxidant Activity of Sulfur and Selenium: A Review of Reactive Oxygen Species Scavenging, Glutathione Peroxidase, and Metal-Binding Antioxidant Mechanisms. Cell Biochemistry and Biophysics, 2009; 55, 1–23.

[10]   ERMAN, J.E.; VITELLO, L.B.; Yeast cytochrome c peroxidase: mechanistic studies *via* protein engineering. Biochimica et Biophysica Acta, 2002; 1597, 193– 220.

[11]   BARCELÓ, A.R.; POMAR, F.; Plant Peroxidases: Versatile catalysts in the synthesis of bioactive natural products. In: Atta-ur-Rahman, Studies in Natural Products Chemistry, 2002; 27, 735-791.

[12]   DORGE, D.R.; DIVI, R.L.; CHURCHWELL, M.I.; Identification of the Colored Guaiacol Oxidation Product Produced by Peroxidases. Analytical Biochemistry, 1997; 250, 10-17.

[13]   AZEVEDO, A.M.; PRAZERES, D.M.F.; CABRAL, J.M.S.; FONSECA, L.P.; Stability of free and immobilised peroxidase in aqueous–organic solvents mixtures. Journal of Molecular Catalysis B: Enzymatic, 2001; 15, 147–153.

[14]   LAURENTI, C.; CLEMENTE, E.; Avaliação da atividade da peroxidase em carambola (*Oxalidacia averrhoa*) em diferentes estádios de maturação. Acta Scientiarum. Agronomy, 2005; 27 (1), 159-163.

[15]   RADIC, S.; RADIC-STOJKOVIC, M.; PEVALEK-KOZLINA, B.; Influence of NaCl and mannitol on peroxidase activity and lipid peroxidation in *Centaurea ragusina* L. roots and shoots. Journal of Plant Physiology, 2006; 163, 1284-1292.

[16]   VOJINOVIC, V.; AZEVEDO, A.M.; MARTINS, V.C.B.; CABRAL, J.M.S., GIBSON, T.D.; FONSECA, L.P.; Assay of H2O2 by HRP catalysed co-oxidation of phenol-4-sulphonic acid and 4-aminoantipyrine: characterisation and optimization. Journal of Molecular Catalysis B: Enzymatic, 2004; 28, 129–135.

[17]   MOLAEI RAD, A.; GHOURCHIAN, H.; MOOSAVI-MOVAHEDI, A.A.; HONG, J.; NAZARI, K.; Spectrophotometric assay for horseradish peroxidase activity based on pyrocatechol–aniline coupling hydrogen donor. Analytical Biochemistry, 200; 7362, 38–43.

[18]   FRICKS, A.T.; SOUZA, D.P.B.; OESTREICHER, *E.G.*; ANTUNES, O.A.C.; GIRARDI, J.S.; OLIVEIRA, D.; DARIVA, C.; Evaluation of radish (*Raphanus sativus L.*) peroxidase activity after high-pressure treatment with carbon dioxide. Journal of Supercritical Fluids, 2006; 38, 347–353.

[19]   CAPEILLÈRE-BLANDIN, C.; Oxidation of guaiacol by myeloperoxidase: a two-electron-oxidized guaiacol transient species as a mediator of NADPH oxidation. Biochemical Journal, 1998; 336, 395-404.

[20]   YÖRÜK, I. H.; DEMIR, H.; EKICI, K.; SAVRAN, A.; Purification and Properties of Catalase from Van Apple (Golden Delicious). Pakistan Journal of Nutrition, 2005; 4 (1), 8-10.

[21]   COLOWICK, S.P.; KAPLAN, N.O., eds., Methods in Enzymology, Vol. II. Academic Press, New York, 1955.

[22]   FAO (FOOD AND AGRICULTURE ORGANIZATION OF THE UNITED NATIONS); Combined Compendium of Food Additive Specifications, Vol.4: Analytical methods, test procedures and laboratory solutions used by and referenced in the food additive specifications. Joint FAO/WHO Expert Committee on Food Additives, Rome, 2006.

Send Orders for Reprints to reprints@benthamscience.net

CHAPTER 3

Methods to Determine Chitinolytic Activity

Rosalie R. R. Coelho*, Juliana P. Rosa, Rodrigo F. Souza and Adriana M. Fróes

Department of General Microbiology, Institute of Microbiology Paulo de Góes, Federal University of Rio de Janeiro, Rio de Janeiro, Brazil

Abstract: In this chapter methods for the selection of chitinolytic microorganisms is discussed, especially those based on the formation of clear hydrolytic zones due to colloidal chitin degradation. Also the main methods for the detection and measurement of chitinolytic activity are presented, including several assays for exochitinase and endochitinase, based on artificial substrates, as well as those based on reducing sugars from chitinous substrates, such as glycol-chitin or colloidal chitin. In addition, methods for the detection of chitinase in polyacrylamide gel electrophoresis are discussed.

Keywords: Bioprospection, endochitinase, exochitinase, chitinolytic activity, colloidal chitin, glycol-chitin, chromogenic substrates, fluorogenic substrates, methylumberlliferyl, Nelson-Somogyi method, p-nitrophenyl, reducing sugars, SDS_PAGE, staining dyes, zymogram, chitinase, β-N-acetyl glucosaminidase, pre-enrichment, chitobiase, N-acetylglucosamine.

INTRODUCTION

Chitin, the second most abundant and renewable biopolymer on earth after cellulose, is composed of a linear chain of β-1,4-linked N-acetyl-D-glucosamine (GlcNAc) units [1]. It is estimated that about 10 gigatons of chitin are produced annually and over 80.000 metric tons of crustaceans marine waste collected [2-6].

Chitin is one of the main components of fungal cell wall, comprising 22 to 40 %.

*Address correspondence to Rosalie R. R. Coelho: Institute of Microbiology, Paulo de Góes at Federal University of Rio de Janeiro, Brazil; Tel: (021) 25626743; Fax: (021) 25608344; E-mail: rosalie@micro.ufrj.br

It is also present in the outer shell of crustaceans, integuments of arthropods, in nematodes eggs, mollusks, and in diatoms, such as *Thalassiosira fluviatilis* [2]. In insects it occurs as a major component of the cuticle, in the peritrophic membrane and also as a protective sleeve of the lining gut. Chitin is insoluble in water, dilute and concentrated alkalis, alcohol and other organic solvents [7]. Though it is a very rigid molecule, due to their physiochemical properties, chitin can be degraded by a system comprised by two enzymes: an endochitinase (EC 3.2.1.14), which cleaves internal linkages randomly throughout the chain of chitin, generating low molecular weight chitooligomers, such as chitotetraose, chitotriose and diacetylchitobiose, and an exochitinase, or β-N-acetylglucosaminidase (EC 3.2.1.52), which cleaves off the non-reducing terminal, releasing units of N-acetylglucosamine, preferring oligomers as chitotriose and chitotetraose. This definition is accepted by several authors [4, 6, 8], including the Nomenclature Committee of the International Union of Biochemistry and Molecular Biology (NC-IUBMB). However, some authors consider another classification, which includes 4 types of chitinase: an endochitinase, similar to that described above, an exochitinase which cleaves from the non reducing terminal, releasing dimmers of N-acetylglucosamine; a chitobiase, which cleaves these N-acetylglucosamine dimmers and an N-acetylglucosaminidase, which cleaves from the non reducing end of chitin, releasing N-acetylglucosamine units [9-12].

Several assay methods, either qualitative or quantitative, can be used to evaluate chitinase activity using different substrates. These include soluble chitinous substrates chemically modified, such as glycol-chitin, or carboxy-methyl chitin, as well as insoluble ones, as tritiated solid chitin, colloidal chitin or chitin covalently coupled to different dyes, as for instance, chitin-azure [13,14]. These different assays have different costs and different performances, and may be used according to the characteristics of the experiment.

In this chapter we will present a simple method to selectively isolate chitinolytic microorganisms, and also discuss some methods to detect and quantify chitinase activity. One of them is based on a reducing sugar assay, the Nelson-Somogyi method, where the concentration of N-acetylglucosamine, the final monomer released as a result of the chitinases' action on chitin, can be measured. One assay based on the use of synthetic substrates, as the fluorogenics methylumbelliferyl

(MU) linked, will also be described. Finally it will be discussed the detection of chitinolytic activity through zymograms, where glycol-chitin, used as substrate, is incorporated to a polyacrylamide gel before electrophoresis occurs. The enzymatic activity is detected after revealing the gel with a fluorescent dye, Calcofluor (Calcofluor White M2R, or CW M2R, Sigma) [15]. This technique also permits the determination of the apparent molecular masses of the enzyme(s).

It is important to have in mind that the quantification of any enzymatic activity is based on the enzyme-substrate reaction, which occurs in certain conditions. Different buffers, with different molarities and pH values, different temperatures and incubation times can be used. For this reason, in each experiment, or in each situation, it is necessary to optimize all reactions, finding the best conditions for obtaining maximal enzymatic activity detection.

Selective Isolation of Chitinolytic Microorganisms

Bioprospection of chitinolytic microbes may become easier if a pre-enrichment of the environmental sample is achieved. For instance, in the case of soil, you can make use of the addition of a crude chitin, such as chitin from crab shells, practical grade (Sigma), or colloidal chitin, for about 30 days before the soil sampling. This step facilitates the posterior isolation of the chitin degraders, as chitin substrate will induce chitinase production enhancing further growth of the chitinolytic microbes.

After obtaining the chitin pre-enriched sample, the isolation can be performed using a dilution plate technique with a selective agar-medium such as a minimal mineral salts solution added of agar and colloidal (1 %) or a crude chitin (1 %) to be used as the sole carbon and nitrogen source. Moreover, another possibility is the use of the dry mycelium of a pathogenic fungus instead of chitin. This is a very interesting approach if you want to study chitinase activity aiming at its application in biocontrol. In this situation you can enhance the chances of finding a chitinolytic strain able to degrade the cell wall of the pathogenic fungus in question, since fungal cell wall are generally rich in chitin. After incubation of the isolation plates in proper conditions, chitinolytic strains can be detected through observation of clear zones around colonies, against the opaque medium containing insoluble chitin or dry mycelium.

Objective

To selectively isolate soil microorganisms capable of producing chitinase.

Principle

This isolation method is based on the capacity of chitin to induce chitinase and enhance the growth of chitinolytic microbes in relation to the whole population of the soil sample.

Material

1. Agar-chitin medium containing a mineral salt solution [16], agar and a source of chitin (*e.g.,* crude chitin, colloidal chitin, fungal dry mycelium, *etc.*);

2. Petri dishes;

3. Soil sample;

4. A source of chitin for soil enrichment (*e.g.,* crude chitin, colloidal chitin, fungal dry mycelium).

Procedure

1. Prepare the pre-enrichment of the soil sample by sieving the collected soil, weighing it and truly mixing it with chitin 1.0 % (w/w) (choose the appropriate source). Pay attention to the humidity of the system, if it is too dry add some distilled water;

2. Incubate for 30 days. Whenever necessary add some extra distilled water to avoid the drying of the soil;

3. Prepare chitin-agar medium in Petri dishes;

4. Perform the dilution plate technique using the chitin pre-enriched soil by pour-plate, for instance;

5. Incubate in the appropriate conditions of temperature and time;

6.   After the incubation period chitinolytic strains grown can be detected on the plates, presenting a clear zone around or below their growth;

7.   Transfer the isolates to a suitable medium and stock appropriately.

Obs.: This experiment should preferably be carried out using triplicates.

Comments

This method is an adaptation of the traditional dilution plate technique for isolation and enumeration of soil microbes. The pre-enrichment of the soil sample and preparation of the cultivation medium with chitin guarantees the selectivity. According to the microbes group you are interested in, you can choose the appropriate conditions of the medium, incubation temperature and pH, *etc.* As chitin is a source of C and N, the medium for this selective isolation can contain only this substrate plus a mineral salt solution, without N salts. This facilitates the growth of solely chitinase producers.

Screening of Chitinase Producing Microorganisms

Chitinase producing microorganisms can be detected by several methods, each one with its own characteristics. The most common ones are those based on the degradation of chitin, in its various forms, but those which use artificial substrates such as the chromogenics have also been cited in literature. It is advisable to use more than one method to guarantee a better chance to select good chitinase producers.

Objective

To select microorganisms capable of producing chitinase based on chitin degradation on agar plate.

Principle

This method is based on the chitin-degrading capacity of microorganisms which produce chitinases. The degradation of the insoluble chitin generates a hydrolysis zone around or below the microbial growth on agar plate.

Material

1. Petri dishes;

2. Agar medium containing chitin[1] (choose the appropriate type [2]) 1.0 % (w/v) and a mineral salts solution (*e.g.,* [16]);

3. Microbial strains of interest.

Procedure

1. Prepare agar medium containing chitin on Petri dishes;

2. Inoculate the microbial strains as spots or as strips on the agar surface;

3. Incubate in adequate conditions of temperature and time [3];

4. Detect the chitinase producing microorganisms by visualizing a hydrolysis zone surrounding or below microbial growth (Fig. **1**).

    Obs.: This test should preferably be carried out using triplicates.

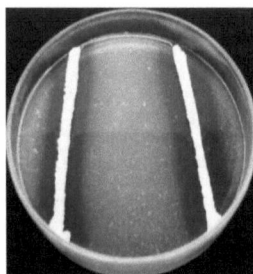

Figure 1: Chytinolytic microorganism growth on a colloidal chitin media and visualization of a clear zone of chitin degradation.

---

[1]Chitin can be used as sole carbon and nitrogen source, but a small amount of a nutrient starter can be added, such as yeast extract or glucose. However, it is necessary to pay attention, since glucose can also induce catabolite repression of microbial chitinases [13].

[2]Besides chitin, colloidal chitin, powder shrimp shell, fungal dry mycelium, or any other type of substrate chitin-rich can be used. Also those where a dye is labeled to chitin, such as Remazol Brilliant Blue R (RBBR) [14] or Remazol Brilliant Violet 5 R (RBV5R) [17,13], also known as chitin-azure [18].

[3]Colonies should not be scored as negative until they have been incubated for several weeks, since depolymerization of particulate chitin may take days or weeks to become visible. For this reason, also, assay plates should be incubated in plastic bags with a damp paper towel to prevent drying during extended incubations.

Comments

As already suggested, alternative substrates may be included in agar media plates to facilitate the visualization of chitin degradation activity. Chitin-azure can be prepared by covalently linking the soluble dye to colloidal chitin [17, 13]. When chitin-azure is depolymerized clear zones appear around the growth, which are easier to visualize than those formed on chitin plates. Chitin-azure should be incorporated into the culture medium at a final concentration of 0.08 %, autoclaved, and incubated for about 2 weeks [13]. The chitin-azure test can also be performed in agar tubes, (Fig. **2**), in an experiment adapted from a cellulase-azure test [19]. In this case the medium is prepared in two stages: first the bottom, containing 5 mL of a basal medium in a semi-solid agar; then, after the solidification of this bottom medium, the superficial layer, containing the same medium plus 1% chitin-azure (500 µL). After inoculation and incubation at 28°C / 14 days, the chitinolytic strains can be detected, as they are able to degrade chitin-azure releasing the dye in the bottom. The stronger the chitinolytic activity, the stronger the colour of the dye in the bottom of the test tube.

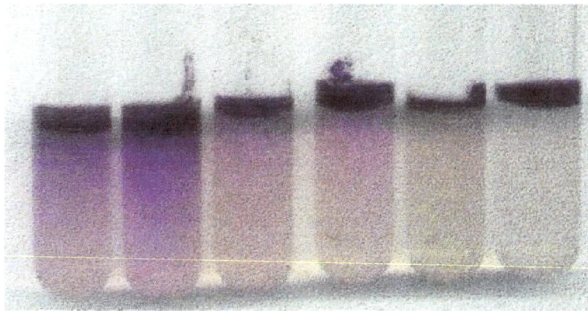

Figure 2: Chytinolytic microorganism growth on a chitin-azure media and visualization of the dye in the botton of the test tube, indicating chitin degradation. Last tube at rigth, negative control.

Ethylene glycol-chitin (EGC) can also be added to agar plates to detect the chitinase producing microorganisms. This soluble form of chitin can be added to the agar medium at 0.4 % plus 0.01 % Trypan Blue. Solutions of Trypan Blue and EGC should be filter-sterilized prior to addition to sterile, molten agar. After culturing, clear zones will appear around growth of chitinolytic microbes [13].

Chitinase activity can also be quantitatively determined by measuring the ratio of the lytic zone around growth colony on agar plates. Different chitinase activities will result in different chitin digestion, giving different diameters of clear zones [20]. However, this methodology has a low sensitivity and the results will depend, also, on the concentration and size of the colloidal chitin particles, the thickness of the media, and the amount and kind of inoculum [14].

Enzymatic Activity Assays

Determination of chitinase activity can be performed in several different ways. For instance, using natural substrates such as chitin, in its soluble (*e.g.,* colloidal chitin) or insoluble forms (*e.g.,* powder chitin) and measuring, spectrophotometrically, the reducing sugars produced. Or using dyes where chitin is covalently linked, such as RBBR [14]. Besides, artificial substrates can also be used, as p-nitrophenyl or methylumbelliferyl linked to oligomers of N-acetyl-β-D-glucosamine (GlcNAc) in its monomeric, dimeric or trimeric form. In those cases, the quantification can be performed through the release of the dyes or the chromogenic compounds (p-nitrophenol) using a spectrophotometer, or using a fluorometer for the fluorogenic (4-methylumbelliferone) compound.

Measurement of Enzymatic Activity Using Colloidal Chitin as Substrate

Objective

To measure the enzymatic activity of chitinase using a natural substrate, colloidal chitin, and detecting the generated reducing sugars using the Nelson-Somogyi method.

Principle

The action of chitinases on the substrate chitin releases chitooligomers, or the monomer, GlcNAc, which are reducing sugars that can be measured by the Nelson-Somogyi technique. In a simplest way, the reaction between reducing sugars and the Nelson-Somogyi reagent is based in the oxidation of the sugars (the C=O groups are oxidized to COOH groups) and reduction of Cu(II) to Cu(I). This is oxidized back to Cu(II) using a colourless hetero-polymolybdate complex that is reduced to give the characteristic blue colour, which is proportional to the amount of reducing sugars generated [21-23].

Material and equipment

1. 20 mL glass tubes;

2. Enzyme sample (*e.g.,* crude extract or supernatant, diluted appropriately, if necessary);

3. Colloidal chitin 0.5 % (w/v) [4];

4. Appropriate incubation buff;

5. Nelson-Somogyi reagent [5];

6. Hot plate stirrer;

7. Ice bath;

8. Water bath with temperature controller;

9. Spectrophotometer (560 nm).

(1)   Colloidal chitin [14] - Sieve, through a # 40 mesh, 10 g of commercial chitin (*e.g.,* Sigma) grounded in a mill. Add the obtained powder to 100 mL of 85 % phosphoric acid and keep in a refrigerator (4°C) for 24 h. Add 2 L of tap water. Separate the gelatinous white material formed by filtration through filter paper. Wash the retained cake with tap water until the pH of the filtrate turns to 6.5. The colloidal chitin obtained will present a soft, pasty consistency, with 90 – 95 % moisture.

(2)   Preparation of the Nelson-Somogyi reagent [24]

Nelson reagent
$(NH_4)6Mo_7 O_{24}. 4H_2O$ – 100 g in 1800 mL of distilled water.
H_2SO_4 concentrated – 84 mL.
$NaAsO_2$ – 12 g dissolved in 100 mL of distilled water.
Stock the solution at 37 °C for 24 – 48 h, in a dark glass.

Somogyi reagent I
Na_2SO_4 anhydrous – 288 g dissolved in 1 L of boiling distilled water.
$KNaC_4 H_4 O_6·4H_2O$ – 24 g.
Na_2CO_3 – 48 g.
$NaHCO_3$ – 32 g.
After dissolving Na_2SO_4 in boiling water, wait for the solution to get colder and then add the other reagents.
Complete to 1.600 mL with distilled water and stock at 27 °C.
Mixture solutions A and B in the proportion 4:1 just before utilization.

Somogyi reagent II
Na_2SO_4 anhydrous – 72 g dissolved in 300 mL of boiling distilled water.
$CuSO_4·5H_2O$ – 8 g.
Stock the solution at 27 °C.

Procedure

1. In a glass tube prepare the reaction mixture (400 μL) with the appropriate buffer (*e.g.,* for *Streptomyces* sp., Tris-HCl pH 7.4, 50 mM), colloidal chitin at a final concentration of 0.5 % and the enzyme sample;

2. Incubate both reaction mixture and reagent blank [6] at the appropriate conditions of time and temperature (*e.g.,* for *Streptomyces* sp., 60 min at 50 °C in a water bath);

3. Immerse the tubes in an ice bath;

4. Centrifuge for few minutes and separate insoluble colloidal chitin;

5. Add 400 μL of the Somogyi reagent I + II (4:1) in the systems;

6. Prepare the standard curve with N-acetylglucosamine [7];

7. Boil supernatant in water for 15 min and place in an ice bath immediately. Wait for a few seconds (15-20 s);

8. Add 400 μL of the Nelson reagent;

9. Stir slowly to release the gas;

10. If the produced color is very intense add distilled water;

11. Measure the produced colour using the spectrophotometer (560 nm) against the appropriate blank [6];

12. Correct the absorbance obtained with the enzyme blank [8] if necessary;

13. Using the N-acetylglucosamine standard curve obtain the angular coefficient of the linear regression;

14. Calculate activity in the original (undiluted) sample by the following equation:

$$IU = \frac{(ABS_{RM} - ABS_{EB}).\, \alpha.\, D}{\Delta T} \tag{1}$$

IU – usually, one enzyme activity international unit is defined as the amount of enzyme required to produce 1 µmol of reducing sugars (N-acetylglucosamine) per minute of reaction under the assay conditions previously described.

α = angular coefficient of the standard linear regression

ABS$_{RM}$ = mean of absorbance of the reaction mixture (after incubation)

ABS$_{EB}$ = mean of absorbance of enzyme blank (without incubation)

D = enzymatic dilution

ΔT = time of incubation (enzyme-substrate reaction) (in min).

Obs.:   *a) When boiling all tubes in water bath be aware to put sufficient water to completely cover the reaction mixture. All samples, controls, blanks and standards should be boiled together.*
b) This test should preferably be carried out using triplicates.

(6)   Reagent blank
-       *Prepare 400 µL of a mixture of the buffer plus colloidal chitin 0.5 % (w/v);*
-       *Continue as the main procedure;*
This reaction corresponds to any color of the reagents, or any color corresponding to the reducing sugar that might be present in the substrate before the enzyme reaction, and that must be subtracted

(7)   N-acetylglucosamine standard curve
-       *Prepare a stock solution of N-acetylglucosamine (100 µg/mL) which can be stored at 20 °C;*
-       *Prepare the standards points, each one containing the final volume of 400 µL of a mixture of buffer plus N-acetylglucosamine at different concentrations;*
-       *Add 400 µL of the Somogyi reagent I+II (4:1);*
-       *Continue as the main procedure;*

As a result different color intensities are generated by the different N-acetylglucosamine concentrations, as shown in (Fig. 3), and the standard curve is constructed using the different values of N-acetylglucosamine concentration (Y axis) versus the correspondent absorbance (X axis) (Fig. 4).

Obs.:   *It is interesting to note that the standard curve can be prepared using concentration units (molarity or µmol mL^{-1}, for instance) or amount units (micromoles or micrograms, as shown in (Fig. 4). If the standard curve is prepared using micromoles units the calculation of the enzymatic activity is as stated in equation 1. However, if it is used any concentration or any amount unit different from micromoles, the calculation of the original activity must take this in account, and the proper transformations must be carried out.*

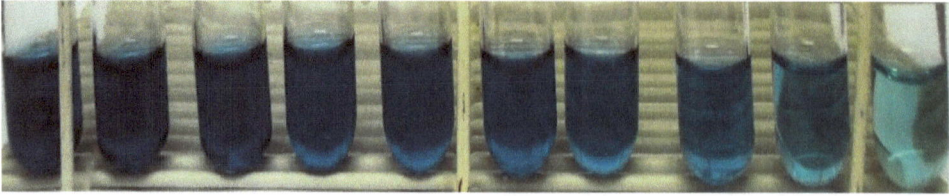

Figure 3: Standard curve of N-acetylglucosamine. The stronger the color means that more N-acetylglucosamine is present in the solution.

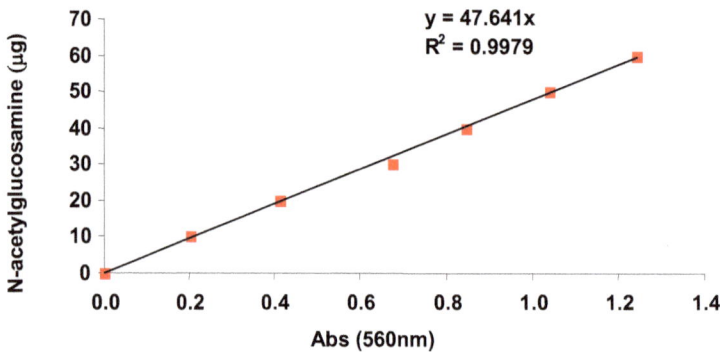

Figure 4: N-acetylglucosamine standard curve indicating the angular coefficient of the standard linear regression ($\alpha = y/x$).

Obs.: This standard curve should preferably be carried out using triplicates

(8) Enzyme blank
- *Prepare the reaction mixture as described for the procedure;*
- *Add 400 µL of the Somogyi reagent I+II (4:1) (this must be done before centrifugation in order to avoid the occurrence of any reaction);*
- *Centrifuge and separate insoluble colloidal chitin;*
- *Add 400 µL of the Nelson reagent;*
- *Continue as the main procedure;*

This reaction corresponds to any reducing sugar that may be present in the enzyme sample (e.g., supernatant or crude enzyme) before the enzyme-substrate reaction occurs, and must be subtracted also. So, it is necessary that the measurement must be performed without the incubation of the system. Enzyme blanks are specially required when the enzyme sample contains a high level of reducing sugars.

Comments

Besides colloidal chitin, commercial chitin, powder shrimp shell, fungal dry mycelium, or any other type of chitin-rich substrate can be used. Also those where a dye is labeled to chitin, such as chitin-azure, can be employed for quantification of chitinase [17, 14].

The amounts and assay conditions used in the procedure here presented can be modified. For instance, different pH buffers, incubation time and temperature are use by different authors in different protocols, depending on the characteristics or the origin of the enzyme sample. Also the sample can be concentrated, if necessary, in AMICON device. In this case, the appropriate corrections must be considered for calculation of the enzymatic activity.

This methodology is commonly used as a cheap option to measure chitinase activity, but it is not as sensitive as those using fluorogenic or chromogenic substrates (see below). So, it is not a good option for the detection and quantification of chitinase activity in samples containing low amounts of the enzyme.

The detection of reducing sugars can also be performed using other methods. The dinitrosalicylic (DNS) method [25], for instance, is a very common assay cited in literature [13], but it is not so accurate and much less sensitive than the Nelson-Somogyi method [26]. Other method very commonly cited is the Schales' procedure [27].

Although all these methods have been largely used, Horn & Eijsink, in 2004 [28] have claimed that they result in an overestimation of reducing end formation, and have proposed a new method which, however, has not been cited often.

Enzymatic Assays Using Artificial Substrates

These methods are based on the utilization of artificial substrates which are linked to chromogenic or fluorogenic compounds for detecting the activity of specific enzymes. The assays are sensitive and allow the detection and quantification of chitinolytic activity. The artificial substrates most commonly used are those which contain one, two or three units of the monomer, N-acetyl-β-D-glucosamine, linked to the fluorogenic 4-methylumbelliferyl, or the chromogenic p-nitrophenyl compound. p-nitrophenol or 4-methylumbelliferone (MU) is released after hydrolysis of the chromogenic or fluorogenic substrates by chitinolytic enzymes presented in the sample. In the present section a procedure using fluorogenic substrates such as methylumbelliferyl (MU) linked, is described.

Objective

To detect and quantify chitinolytic activity using methylumbelliferyl linked substrates and measuring the absorbance of the methylumbelliferone released after the enzyme reaction, performed in a micro titer 96-well plate.

Principle

Substrates consisting of methylumbelliferyl attached to chitin analog monomer or oligomers compound can be used to quantify chitinases. Exochitinase activity can be detected using 4-methylumbelliferyl-N-acetyl-β-D-glucosamine (4-MU-GlcNAc) and 4-methylumbelliferyl-β-D-N,N'-diacetylchitobioside[4-MU-(GlcNAc)$_2$], whereas the detection of the endochitinase activity can be obtained when using 4-methylumbelliferyl-β-D-N,N''-triacetylchitotriose [4-MU-(GluNAc)$_3$] or 4-methylumbelliferyl-β-D-N,N'''-tetracetylchitotetraose [4-MU-(GluNAc)$_4$]. The increase in fluorescence caused by the fluorogenic substrate fission and release of 4-MU molecule can be measured in a fluorometer.

Material and Equipment

1. Black 96-well microplates;

2. Enzyme sample (*e.g.*, crude extract or supernatant, concentrated appropriately if necessary);

3. Appropriate incubation buffer;

4. 4-MU-GlcNAc 50 μM;

5. 4-MU-(GlcNAc)$_2$ 50 μM;

6. 4-MU-(GluNAc)$_3$ 50 μM;

7. 4-MU-(GluNAc)$_4$ 50 μM;

8. Water bath with temperature controller;

9. NaOH 0.02M;

10. Fluorescence microplate reader (excitation 355 nm and emission 460 nm).

Procedure

1. In a microplate well, add 100 μL of the enzyme sample, 50 μL of appropriated buffer (*e.g.,* for streptomycetes, Tris-HCl 50 mM pH 7.4)

and 50 µL of the substrate to be tested, 4-MU-GlcNAc 50 µM, [4-MU-(GlcNAc)$_2$] 50 µM, [4-MU-(GluNAc)$_3$] 50 µM or [4-MU-(GluNAc)$_4$] 50 µM;

2. Incubate the reaction mixture at chitinase optimal time and temperature (*e.g.,* for *Streptomyces* sp., 60 min at 50 °C);

3. Prepare the standard curve with 4-methylumbelliferone [9];

4. Prepare the enzyme blank [10];

5. Stop the enzyme reaction adding 30 µL of NaOH 0.02 M;

6. Measure the fluorescence produced using the fluorescence microplate reader;

7. Using the standard curve of 4-methylumbelliferone obtain the angular coefficient (α) of the linear regression;

8. Calculate activity in the original (undiluted) sample by the following equation:

$$IU = \frac{\left(ABS_{RM} - ABS_{EB}\right) . \, \alpha. \, D}{\Delta T} \qquad (2)$$

IU = usually, one international unit of enzymatic activity corresponds to the amount of enzyme required to produce 1 µmol of MU per min of reaction under the assay conditions previously described.

α = angular coefficient of the standard linear regression

ABS$_{RM}$ = mean of absorbance of the reaction mixture (after incubation)

ABS$_{EB}$ = mean of absorbance of enzyme blank (without incubation)

D = dilution of the enzyme

ΔT = time of incubation (enzyme-substrate reaction) (in min).

Obs.:

- *Depending on the level of enzymatic activity the units can be changed. For instance, one enzymatic activity can be considered the amount of enzyme required to produce 1 μmol of MU in one hour of reaction under the assay conditions. In this case this will not be the International Unit.*

- *This test should preferably be carried out using triplicates.*

[9] Methylumbelliferone standard curve

- *Prepare a stock solution of 4-MU 500 μM (e.g., Sigma) in the appropriate buffer and store it at -20 °C;*
- *Prepare the standard points, each one containing the final volume of 200 μL and a mixture of buffer plus the different amounts of 4-MU (e.g., 12 points, from 0.1 to 10 μmoles of 4-MU);*
- *Continue as the main procedure;*

The standard curve is usually constructed using the different values of 4-methylumbelliferone concentration or amount of the standard points (Y axis) vs. the correspondent absorbance (X axis) (Fig. 5).

Obs.: *a) It is interesting to note that the standard curve can be prepared using concentration units (molarity or μmol mL^{-1}, for instance) or amount units (micromoles or micrograms, as shown in Fig. 5). If the standard curve is prepared using micromoles units the calculation of the enzymatic activity is as specified in equation 2. However, if it is used any concentration or any amount unit different from micromoles, the calculation of the original activity must take this in account, and the proper transformations must be carried out.*

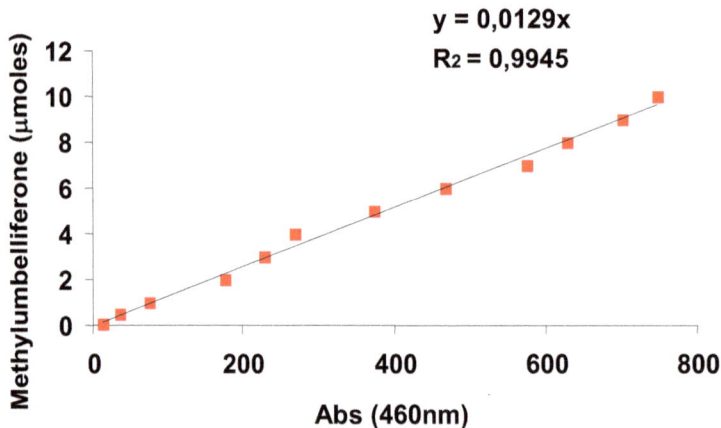

Figure 5: Methylumbelliferone (4-MU) standard curve indicating the angular coefficient of the standard linear regression ($\alpha = y/x$).

b) *The standard curve should preferably be carried out using triplicates.*
c) *The amounts and concentrations used in the enzymatic reaction, as well as in the standard curve are variable, according to the characteristics of the enzyme.*

[10] Enzyme blank.

- *Prepare the reaction mixture (100 μL of the substrate, 50 μL of the enzyme sample and 50 μL of appropriated buffer);*
- *Add 30 μL NaOH 0.02 M;*
- *Continue as the main procedure;*

This reaction corresponds to any MU that may be present in the reaction mixture due to a possible substrate self degradation or due to fast reaction between the substrate and the enzyme sample, before incubation (while the assay is being prepared).

Comments

Fluorogenic and chromogenic artificial substrates are widely used for studying chitinase activity and are considered the most sensitive methods used for chitinase assay. Although being readily available commercially, they are very expensive, which make them suitable to study specificity of chitinases more than to select chitinolytic strains.

The choice of the substrate will depend on the chitinase activity type to be detected. The trimeric and the tetrameric ones can detect an endochitinase activity; whereas the dimeric an exochitinase activity and the monomeric the chitobiase activity. However, as indicated earlier in this section, different authors use different criteria, and sometimes chitobiase is also considered an exochitinase activity.

Considering the chromogenic substrates, the following oligomers of N-acetyl-β-D-glucosamine (GlcNAc) can be used as monomeric, dimeric and trimeric substrates: p-nitrophenyl-N-acetyl-β-D-glucosaminide (pNP-GlcNAc), p-nitrophenyl-β-D-N,N'-diacetylchitobiose [pNP-(GlcNAc)$_2$], and p-nitrophenyl-β-D-N,N',N''-triacetylchitotriose [pNP-(GlcNAc)$_3$], respectively [9,29,30].

Obviously, in the case of the monomeric substrate, any enzymatic reaction will release 4-methylumbelliferone or p-nitrophenol, respectively, and, therefore, the enzymatic activity will be more rapidly detected than with the dimeric and trimeric substrates, which need two and three cleavages for the release of the p-nitrophenol or 4-MU, respectively [31].

It is interesting to mention that artificial substrates with substituted aglyconic moieties, such as p-nitrophenyl and -methylumbelliferyl, are more prone to the hydrolysis by the enzyme than the natural substrates having only sugar residues, such as chitobiose [2] and this must be taken in consideration when choosing the method for studying chitinase activity.

Measurement of Enzymatic Activity Using Dyes Linked to Chitin

Several methods are described in literature for measuring chitinase activity using a substrate where a dye is covalently linked to chitin. The chitin can be used in an insoluble form, such as powder chitin [32, 33]or in a soluble form, such as carboxymethyl-chitin (CM-chitin) [34-37] or colloidal chitin [14]. The dyes RBV5E [34-37] or RBBR [14] are the most cited ones.

These assays are based in the release of the soluble dye after the enzymatic action. The substrate can be easily separated from the system by a precipitation with HCl and centrifugation [35-37] or only a centrifugation [33]. The measurement of the increase of the absorbance will correspond to the increase in the release of the dye, which, in turn, is a measurement of the enzymatic activity.

These methods can be used in a conventional from, but they have also been adapted to microplate assays, using micro titer plates of 96 wells [34, 37]. This modification uses small volumes of samples and reagents are rapid and very useful for a large number of samples to be tested.

Chitinases Activity in Polyacrylamide Gel: Zymogram

Methods for the detection of chitinase activity using electrophoresis have been available since 1989, when Trudel and Asselin [15] presented a methodology which uses denaturing polyacrylamide gel electrophoresis (PAGE) for the separation of the different enzymes. By adding glycol-chitin into the gel, dark zones can be visualized at the position of the chitinase activities using Calcofluor fluorescent staining and a UV transiluminator.

Objective

Study the profile of chitinases using a sodium-dodecyl-sulphate polyacrylamide gel electrophoresis (SDS-PAGE) system containing glycol-chitin, and determine the apparent molecular masses of each of the enzymes.

Principle

This methodology is based on the substrate degradation (usually glycol-chitin) by the chitinases present in the sample, which can be, for instance, the crude extract obtained from cultivation of chitinolytic microorganism. After an electrophoretic run in polyacrylamide gel, and consequent diffusion of the chitinases, the gel is incubated under appropriate conditions to allow the enzyme reaction to occur, with the consequent cleavage of the substrate [38]. After substrate degradation, Calcofluor fluorescent dye is added and the hydrolysis bands formed can be visualized by UV light. The dye binds only to the undigested chitin, specifically to linear β-(1,4)-glucosidically linked units of N-acetylglucosamine [39]. After binding to chitin, the flourochrome highlights and emits a blue light when exposed to UV. During chitin degradation to individual subunits, the fluorescence is lost and this is indicated by a dark unstained band against a fluorescent background [40]. The determination of the apparent molecular mass is possible by using in parallel, a pre-stained molecular mass standard

Material and Equipment

1. Running gel;

2. Stacking gel;

3. Enzyme sample (if necessary, concentrate the enzyme extract, using for instance an AMICON system);

4. Pre-stained molecular mass standard (*i.e.*: 6.0 – 181.8 kDa - Invitrogen);

5. Electrophoresis system;

6. Triton X-100 (1 % v/v in water);

7. Orbital shaker;

8. Appropriate incubation buffer;

9. Water bath with temperature controller;

10. Calcofluor (0.01 % w/v);

11. Distilled water;

12. Transiluminator with UV light.

Procedure

1. Prepare the running gel: a polyacrylamide gel (usually 10 %) containing glycol-chitin 1 % (w/v) [11]. Transfer the solution to the appropriate device;

2. Wait 30-40 min and check polymerization;

3. Prepare the stacking gel: a polyacrylamide gel (4 %) [12];

4. Accommodate the stacking gel above the running gel;

5. Insert the comb into the stacking gel to generate the slots;

6. Wait 30-40 min and check for polymerization;

7. Remove the comb and release the gel base;

8. Place the gel into the electrophoresis cube, add the running buffer (cold) [13] and leave at 4 °C (until the time of the run);

9. Add in a slot, up to 50 μL (or other volume according to the size of the comb and the gel) of the sample containing enzyme plus sample buffer [14] (the exact volume of the enzyme will depend on the enzyme activity and/or protein content);

10. Add, in a separate slot, around 5 μL of pre-stained molecular mass standard;

11. Close the system and run the gel at 100 V and 30 mA until the standards are almost at the end of the gel (this system must be maintained under refrigeration);

12. Transfer the gel to a recipient containing the renaturing solution (Triton X-100);

13. Keep the recipient away from light and incubate for 1 h;

14. Wash the gel with distilled water to remove the renaturing solution;

15. Incubate the gel with an appropriated buffer at optimal time and temperature (*e.g.,* for streptomycetes, Tris-HCl 50 mM pH 7.4 for 1 h at 50 °C);

16. Wash the gel with distilled water;

17. Stain the gel with Calcofluor 0.01 % (w/v) for about 10 min at room temperature, in the dark and in orbital agitation [15];

18. Distain the gel with distilled water for about 4 h at room temperature, in the dark;

19. With the aid of a transiluminator UV light, observe the hydrolysis zones indicating enzymatic activity;

20. Calculate molecular mass [16].

Obs.: The concentration of the polyacrylamide gel, as well as the voltage of the run can be modified, if necessary, as to obtain better results.

[11] Preparing the running gel:

Solutions	Concentration	Volume (μL)
Distilled water	-	2500
Bis-acrylamide	30 % (w/v)	3330
Tris-HCl buffer (pH 8.8)	1.5 M	2500
Glycol-chitin	1.0 % (w/v)	460
Glycerol	-	1100
Ammonium persulphate (APS)	10 % (w/v)	50
Tetramethylethylenediamine (TEMED)	-	5

Obs: *a) Preparation of the Tris-HCl buffer 1.5 M (pH 8.8) – Tris (36.3 g) + HCl 4 M (until pH 8.8). Complete to 200 mL with distilled water.*

b) Tetramethylethylenediamine (TEMED) should be added at the end, just before transferring the solution to the appropriate device, because when it gets in contact with ammonium persulphate (APS) both will catalyze, together, the acrylamide polymerization in the running gel.

[12] Preparing the stacking gel:

Solutions	Concentration	Volume (μL)
Distilled water	-	3000
Bis-acrylamide	30 % (w/v)	670
Tris-HCl buffer (pH 6.8)	0.5 M	1250
Ammonium persulphate (APS)	10 % (w/v)	25
Tetramethylethylenediamine (TEMED)	-	2,5

Obs.: *a) Bis-acrilamide – Acrilamide (60 g) + Bis – Acrilamide (1.6 g). Complete to 200 mL with distilled water.*
b) Tris-HCl buffer 0.5 M (pH 6.8) – Tris (3 g) + HCl 4 N (until pH 6.8). Complete to 200 mL with distilled water.
c) Tetramethylethylenediamine (TEMED) should be added at the end, just before transferring the solution to the appropriate device, because when it gets in contact with ammonium persulphate (APS) both will catalyze, together, the acrylamide polymerization in the running gel.

[13] Preparing the running buffer:

Solutions	Amount
SDS	1.0 g
Glycine	14.4 g
Tris	3.03 g
Distilled water	q.s. 1000 mL

[14] Preparing the sample buffer:

Solutions	Amount
SDS 10 % (w/v)	4.0 mL
Glycerol	2.0 mL
Tris-HCl 0.5 M pH 6.8	2.5 mL
β-mercaptoethanol 2 % v/v or Dithiotreitol (DTT)	0.2 mL/ 0.31 g
Bromophenol Blue 0.02 % (w/v)	0.2 mg
Distilled water	q.s. 10 mL

[15] When using a pre-stained standard molecular mass the position of the migration of each protein can disappear after Calcofluor staining. For this reason you may mark these positions in the gel before staining.

[16] Calculating the apparent molecular masses of chitinases - The use of molecular mass standards turns possible the deduction of an apparent molecular mass for chitinases forms visualized on electrophoresis gel after revelation using Calcofluor. To achieve this, a standard line using a logarithmic curve must be built. The curve shows a correlation between molecular masses and the distance, measured in centimeters, corresponding to the running of the molecular mass standard in gel from the slot where it was applied. Table **1** shows an example using molecular mass standards and the distance they migrated in an electrophoresis gel. Using the formula of the standard curve, obtained from the standards of molecular masses (Fig. **6**), the apparent molecular masses of the chitinases found in the gel can be deduced. The distance between the band and the beginning of the running gel is used as the independent variable of the formula, resulting in an apparent molecular mass of the band.

Table 1: Standard curve: example of molecular masses from standards proteins and the distances they migrated in an electrophoresis gel.

Standard Molecular Masses (kDa)	Distance (cm)
181.8	0.7
115.5	1.4
82.2	2.1
64.2	2.4
48.8	2.8
37.1	3.4
25.9	4.3
14.8	5.3

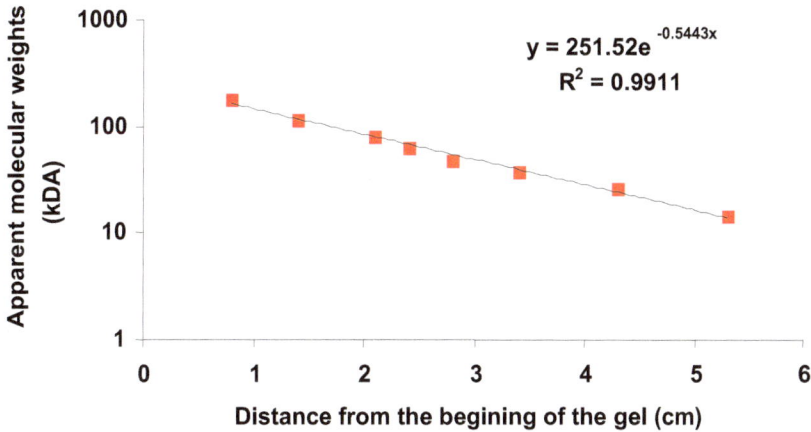

$$y = 251.52e^{-0.5443x}$$
$$R^2 = 0.9911$$

Figure 6: Standard curve using apparent molecular mass of standards related to their migration in electrophoresis gel.

Comments

The substrate glycol-chitin exhibits high affinity toward Calcofluor. So, as shown in (Fig. **7**), the hydrolysis can be visualized as non-fluorescent dark bands in contrast with the fluorescent intact glycol-chitin by UV illumination. Moreover, it is possible to estimate the apparent molecular mass of the chitinolytic enzymes by comparing the enzyme profile with the profile of the standard molecular mass.

A (kDa)   B (kDa)

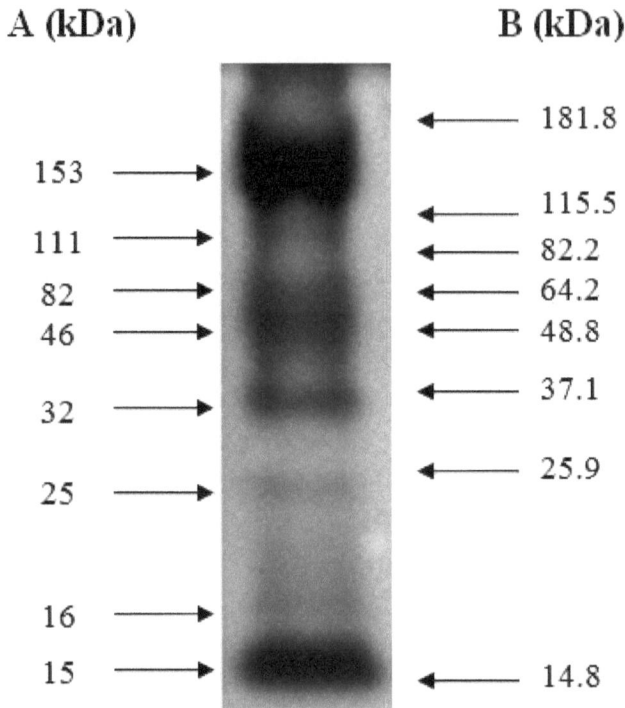

Figure 7: Chitinases profile in a polyacrylamide gel containing 1% glycol-chitin obtained from the crude extract 10-fold concentrated of a streptomycete strain (A), and the position of the bands obtained from the molecular mass standard (B).

Chitinolytic activity can also be detected on polyacrylamide gel electrophoresis by using fluorescent substrates, 4-MU-GlcNAc, 4-MU-(GlcNAc)$_2$ and 4-MU-(GlcNAc)$_3$, as described by [41]. Chitinolytic enzymes appear as fluorescent bands under UV light because of enzymatic hydrolysis of the fluorescent substance 4-MU from the N-acetylglucosamine mono- and oligosaccharides. The molecular mass of the renatured chitinases can be estimated by using pre-stained molecular mass standards.

The method for detecting chitinolytic activity using chitin-containing polyacrylamide gel electrophoresis (PAGE) method with Calcofluor stain is sensitive and reliable. However, in some cases, simultaneous visualization of multiple chitinases in a complex mixture is difficult. So, a silver stain was introduced in a modified method, which successfully overcame the narrow dynamic range problem of the Cw-M2R, but was still time consuming and complex [42]. More recently the method for detecting chitinolytic enzymes in PAGE using modified Comassie Brilliant Blue G 250 (CBB-G250) staining was presented and demonstrated to be more sensitive and less complicated than the conventional Cw-M2R staining. The staining formed achromatic zones at the locations of the migrated enzyme [43].

CONCLUSIONS AND PERSPECTIVES

In this chapter several methods for detecting end/or measuring chitinolytic activity were presented. Methods that use colloidal chitin as a substrate have one serious disadvantage: colloidal chitin is not always dispersed homogeneously in the substrate. This decreases sensitivity and reliability for some of the techniques [44]. As to the chromogenic substrates, despite they are readily available commercially and are easy to obtain, they are not specific for chitinases, and may also serve as substrate for lysozymes and exo-β-N-acetylglucosaminidases [45]. For instance, the compound p-nitrophenyl-β-D-N,N'-diacetylchitobiose was originally synthesized as a potential chromogenic substrate for lysozyme [46]. However, its structure suggests that it might be an appropriate substrate for detection and quantification of chitinase. As to glycol-chitin, this is a soluble modified form of chitin, which has become a very useful substrate for activity staining. However, *in situ* gel activity staining method has the disadvantage that the gel cannot be used further for protein staining and mobility of chitinase in the gel are disturbed because of the presence of polysaccharide in the gel.

It is very important to mention that all methodologies here described have to be optimized for the chitinase system which is being studied. For instance, chitinase from different origins have different characteristics as best pH for enzyme activity. Other important factor that might affect chitinase activity is incubation temperature.

Another important issue is in respect to the use of triplicates in all experiments and the use of appropriate blanks for each reaction. Also it is worth reminding the importance of considering the dilution factor for the calculations of the enzymatic activity.

The study of chitinases and chitinolytic microorganisms are becoming more frequent in literature. Several other methodologies than those presented in this chapter have been described for detecting and measuring chitinolytic activities. For instance, detection of chitinase activity can also be performed onto a solid plate method after polyacrylamide gel electrophoresis under native or denaturing conditions. After running polyacrylamide gel electrophoresis, the gel is transferred to a chitin agar plate containing different dyes for the activity staining. After incubation, hydrolysis zones (bands) of chitinase are visible by daylight or UV light depending of the staining dye used. This method is a simple, rapid, reproducible, sensitive, user-friendly, reliable and cost effective activity staining for chitinase detection in native and denatured polyacrylamide gel [40].

Another gel-diffusion assay for detecting and measuring the activity of chitinase from crude extracts was developed by Zou, Nonogaki & Welbaum in 2002 [38]. In this assay, chitinase diffuses from a small circular well cut in an agarose or agar gel containing the substrate glycol-chitin in a Petri dish. The fluorescent dye Calcofluor is added after an incubation period, and binds only to undigested chitin, leaving an unstained, dark circular zone around each well. Sample activities can be determined from linear regression of log standard enzyme concentration *vs.* the zone diameter of internal standards on each Petri dish used for a diffusion assay. The method is a simple gel-diffusion assay which undergraduate students with minimal training can use to rapidly and inexpensively detect and measure chitinase activity without extensive training, radioisotopes, or expensive specialty equipment. A similar method is the computer-facilitated analysis of the digital pictures of the chitinase assay using a gel-diffusion in Petri dishes. The result is a well-defined dark area on a fluorescent background viewed under UV light transiluminator easily digitally analyzed. This method is reproducible and reliable. Most important, it is fast and allows analysis of large number of samples with minimum effort [47].

ACKNOWLEDGEMENTS

This work was supported by Fundação Carlos Chagas Filho de Amparo a Pesquisa do Estado do Rio de Janeiro (FAPERJ), Conselho Nacional de

Desenvolvimento Científico e Tecnológico (MCT/CNPq) and Coordenação de Aperfeiçoamento de Pessoal do Ensino Superior (CAPES).

CONFLICT OF INTEREST

The authors confirm that this chapter content has no conflict of interest.

REFERENCES

[1]   NEERAJA, C.; ANIL, K.; PURUSHOTHAM, P.; SUMA, K.; SARMA, P. V. S. R. N.; MOERSCHBACHE, B. M.; PODILE, A. R. Biotechnological approaches to develop bacterial chitinases as a bioshield against fungal diseases of plants. Crit Rev Biotechnol 2010; 30: 231-241.

[2]   PATIL, R. S.; GHOMADE, V.; DESHPANDE, M. V. Chitinolytic enzymes: an exploration. Enz Microbiol Technol 2000; 26: 473-83.

[3]   SCHREMPF, H. Recognition and degradation of chitin by streptomycetes. Antonie Van Leeuwenhoek, 2001; 79: 285-289.

[4]   MERZENDORFER, H.; ZIMOCH, L. Chitin metabolism in insects: structure, function and regulation of chitin synthases and chitinases. J Exp Biol 2003; 206: 4393-4412.

[5]   KASPRZENWSKA, A. Plant chitinases - Regulation and function. Cel Mol Biol Lett 2003; 8: 809-24.

[6]   GOHEL, V.; SINGH, A.; VIMAL, M.; ASHWNI, P.; CHATPAR, H. S. Bioprospecting and antifungal potential of chitinolytic microorganisms. African J Biotechnol 2006; 5: 54-72.

[7]   NAWANI, N. N.; KAPANDIS, B. P. Production dynamics and characterization of chitinolytic system of *Streptomyces* sp. NK1057, a well equipped chitin degrader. World J Microbiol Biotechnol 2004; 20: 487-494.

[8]   KUBIEC, C. P.; MACH, R. L.; PETERBAUER, C. K.; LORITO, M. *Trichoderma*: From genes to biocontrol. J Plant Pathol 2001; 83: 11-23.

[9]   HARMAN, G. E.; HAYES, C. K.; LORITO, M.; BROADWAY, R. M.; DI PIETRO, A.; DETERBAUER, C.; TRONSMO, A. Chitinolytic enzymes of *Trichoderma harzianum*: purification of chitobiosidase and endochitinase. Phytopathol 1993; 83: 313–18.

[10]   ULHOA, C. J.; PEBERDY, J. F. Effect of carbon sources on chitobiase production by *Trichoderma harzianum*. Mycol Res 1993; 97: 45-8.

[11]   WANG, S.; MOYNE, A.; THOTTAPPILLY, G.; WU, S.; LOCY, R. D.; SINGH, N. K. Purification and characterization of a *Bacillus cereus* chitinase. Enzyme Microb Technol 2001; 28: 492-8.

[12]   MATSUMOTO, Y., CASTAÑEDA, G.S., REVAH, S., SHIRAI, K. Production of β-N-acetylhexosaminidase of *Verticillum lecanii* by solid state and submerged fermentations utilizing shrimp waste silage as substrate and inducer. Process Biochem 2004; 39: 665-71.

[13]   HOWARD, M. B., EKBORG, N. A., WEINER, R. M., HUTCHESON, S. W. Detection and characterization of chitinases and chitin-modifying enzymes. J Ind Microbiol Biotechnol 2003; 30: 627-635.

[14]   RAMÍREZ, M. G., ROJAS AVELIZAPA, L. I., ROJAS AVELIZAPA, N. G., CRUZ CAMARILLO, C. Colloidal chitin stained with Remazol Brilliant Blue R, a useful substrate to select chitinolytic microorganisms and to evaluate chitinases. J Microbiol Methods 2004; 56: 213– 219.

[15]   TRUDEL, J.; ASSELIN, A. Detection of chitinase activity after polyacrylamide gel electrophoresis. Anal Biochem 1989; 178: 362–66.

[16]   HSU, S. C.; LOCKWOOD, J. L. Powdered chitin agar as a selective medium for enumeration of actinomycetes in water and soil. Appl Microbiol 1975; 29: 422-426.

[17]   SOMMERS, P.; YAO, R.; DOOLIN, L.,; MCGOWAN, M.; FUKUDA, D.; MYNDERSE J. Method for the detection and quantitation of chitinase inhibitors in fermentation broths; isolation and insect life cycle effect of A82516. J Antibiot 1987; 40:1751–1756.

[18]   GUO, S. H.; CHEN, J. K.; LEE, W. C. Purification and characterization of extracellular chitinase from *Aeromonas schubertii*. Enzyme Microb Technol 2004; 35: 550-556.

[19]   PLANT, J. E., ATTWELL, R.W., SMITH, C.A. A semi-micro quantitative assay for cellulolytic activity in microorganisms. J Microbiol Methods 1988; 7:259-263.

[20]   ROJAS, A. L.; CRUZ, C. R.; GUERRERO, M. I.; RODRÍGUEZ VAZQUEZ, R.; IBARRA, J. E. Selection and characterization of a proteo-chitinolytic strain of *Bacillus thuringiensis*, able to grow in shrimp waste media. World J Microbiol Biotechnol 1999; 15: 299– 308.

[21]   NELSON, N. A photometric adaptation of the Somogyi method for the determination of glucose. J Biol Chem 1944; 153: 375-380.

[22]   SOMOGYI, M. A new reagent for the determination of sugars. J Biol Chem 1945; 160: 61-68.

[23]   SOMOGYI, M. Notes on sugar determination. J Biol Chem 1952; 195: 19-23.

[24]   WOOD, T. M.; BHAT, K. M. Methods for measuring cellulase activities. In: Wood, W. A. and Kellog, S. T. (ed.). Methods in Enzymology. San Diego: Academic Press 1988; 160, ch 9.

[25]   MILLER, G. L. Use of dinitrosalicylic acid reagent for determination of reducing sugar. Anal Chem 1959; 31: 426–428.

[26]   BIELY, P.; PUCHART, V. Recent progress in the assays of xylanolytic enzymes. J Sci Food Agric 2006; 86:1636-47.

[27]   IMOTO, T.;YAGISHITA, K. A simple activity measurement of lysozyme. Agric Ecosyst Environ 1971; 35: 1154– 1156.

[28]   HORN, S. J.;EIJSINK, V. G. H. A reliable reducing end assay for chito-oligosaccharides. Carbohydr Polym 2004; 56: 35–39.

[29]   DE LA CRUZ, J.; HIDALGO-GALLEGO, A.; LORA, J. M.; BENITEZ, T.; PINTOR-TOR, J. A.; LLOBELL, A. Isolation and characterization of three chitinases from *Trichoderma havzianum*. Eur J Biochem 1992; 206: 859-67.

[30]   CHERNIN, L.; ISMAILOV, Z.; HARAN, S.; CHET, I. Chitinolytic *Enterobacter agglomerans* antagonistic to fungal plant pathogens. Appl Environ Microbiol 1995; 61: 1720-6.

[31]   LORITO, M.; HAYES, C. K.; DI PIETRO, A.; WOO, S. L.; HARMAN, G. E. Purification, characterization and synergistic activity of a glucan 1,3-β-glucosidase and N-acetyl-β-glucosaminidase from *Trichoderma harzianum*. Phytopathol 1994; 84: 398–405.

[32]   EVRALL, C. C.; ATTWEL, R. W.; SMITH, C. A. A semi-micro quantitative assay for determination of chitinolytic activity in microorganisms. J Microbiol Methods 1990; 12: 138-187.

[33]   SKINDERSOE, M. E.; ALHEDE, M.; PHIPPS, R.; YANG, L.; JENSEN, P. O.; RASMUSSEN, T. B.; BJARNSHOLT, T.; TOLKER-NIELSEN, T.; HOIBY, N.; GIVSKOV, M. Effects of antibiotics on quorum sensing in *Pseudomonas aeruginosa*. Antimicrob Agents Chemother 2008; 52: 3648-43.

[34]   WIRTH, S. J.; WOLF, G. A. Dye-labelled subastrates for the assay and detection of chitinase and lysozyme activity. J Microbiol Methods 1990; 12: 197-205.

[35]   LERNER, D.R.; RAIKHEL, N. V. The gene of stinging nettle lectin (*Urtica dioica agglutinin*) encodes both a lectin and a chitinase. J Biol Chem 1992; 267: 11085-11091.

[36]   FOLDERS, J.; ALGRA, J.; ROELOFS, M. S.; VAN LOON, L. C.; TOMWEIGHTEN, J.; BITTER, W. Characterization of *Pseudomonas aeruginosa* chitinase, a gradually secreted protein. J Bacteriol 2001; 183: 7044-7052.

[37]   BRAND, T.; BEATRIX, W. A. Improved assays for assessment of enzyme activity in nutrient solutions closed irrigation systems. Scientia Horticulturae 2003; 98: 91-97.

[38]   ZOU, X.; NONOGAKI, H.; WELBAUM, G. E. A gel diffusion assay for visualization and quantification of chitinase activity. Mol Biotechnol 2002; 22: 19-23.

[39]   MAEA, H.; ISHIDA, N. Specificity of binding of hexapyranosyl polysaccharides with fluorescent brightener. J Biochem 1965; 62: 276–278.

[40]   GOHEL, V.; VYAS, P.; CHHATPAR, H. S. Activity staining method of chitinase on chitin agar plate through polyacrylamide gel electrophoresis. African J Biotechnol 2005; 4: 87-90.

[41]   TRONSMO, A.; HARMAN, G. E. Detection and quantification of N-acetyl-β-D-glucosaminidase, chitobiosidase, and endochitinase in solutions and on gels. Anal Biochem 1993; 208: 74–79.

[42]   MAREK, S. M.; ROBERT, C. A.; BEUSELINCK, P. R.; KARR, A. L. Silver stain detection of chitinolytic enzymes after polyacrylamide gel electrophoresis. Anal Biochem 1995; 230: 184–185.

[43]   LIAU, C. Y.; LIN, C. S. A modified comassie brilliant Blue G 250 staining method for the detection of chitinase activity and molecular weight after polyacrylamide gel electrophoresis. J Biosci Bioeng 2008; 106: 111–13.

[44]   NITODA. T.; KURUMATANI, H.; KANZAKI, H.; KAWAZU, K. Improved bioassay method for *Spodoptera litura* chitinase inhibitors using a colloidal chitin powder with a uniform particle size as substrate. Pestic Sci 1999; 55: 563– 5.

[45]   ROBERTS, W. K.; SELITRENNIKOFF, C. P. Plant and bacterial chitinases differ in antifungal activity. J Gen Microbiol 1988; 134: 169-176.

[46]   OSAWA, T. Lysozyme substrates. Synthesis of p-nitrophenyl-2-acetamido-4-O-(2-acetamido-2-deoxy-β-D-glucopyranosyl)-2-deoxy-β-D-glucopyranoside and its β-D-(1→6) isomer. Carbohydr Res 1966; 1: 435-43.

[47]   VELASQUEZ, L.; HAMMERSCHMIDT, R. J. Development of a method for the detection and quantification of total chitinase activity by digital analysis. Microbiol Methods 2004; 59: 7–14.

Send Orders for Reprints to reprints@benthamscince.net

Methods to Determine Enzymatic Activity, 2013, 68-99

CHAPTER 4

Cellulase Activity Assays: A Practical Guide

Cristiane Sanchez Farinas[1], Mônica Caramez Triches Damaso[2] and Sonia Couri[3,*]

[1]*Embrapa Intrumentation, São Carlos-SP, Brazil;* [2]*Embrapa Agroenergy, Brasília-DF, Brazil and* [3]*Federal Institute of Science, Education and Technology of Rio de Janeiro, Rio de Janeiro-RJ, Brazil*

Abstract: Cellulase enzymes constitute a complex of glycoside hydrolases that are secreted by microorganisms, plants and some animals. These enzymes act on the hydrolysis of β-1,4–glucosidic bonds of the cellulose structure, the major polymer present on biomass. Recent attention has been given on cellulases use in the bioconversion process of lignocellulosic materials into bioethanol and biobased products, within the biorefinery concept. Besides that, cellulases have been largely applied at different industrial areas like textile, food, brewery and wine, animal feed, pulp and paper industries. This chapter will describe some current state of knowledge related to cellulase assays using soluble and insoluble substrates, focusing on the need to standardized methods in order to allow a comparison among different enzyme preparations.

Keywords: Enzyme, cellulase, polysaccharides, enzyme assays, activity, endoglucanase, β-glucosidases, exoglucanases, enzyme action, hydrolysis, DNS, glucose, cellobiose, cellulose, *Trichoderma reesei*, *Aspergillus niger,* biofuels, bioconversion, reducing sugar, glycosidic bond.

INTRODUCTION

Cellulases are enzymes which degrade the insoluble polymer cellulose. In order to perform these task bacteria, fungi, plants and insects have developed a variety of different systems with multiple cellulases [1]. These enzymes have a synergistic action when degrading the polymeric chain. The most accepted mechanism describes the action of three classes: endoglucanases, exoglucanases and β-glucosidases (Fig. **1**). Endoglucanases hydrolyze accessible intramolecular β-1,4-glucosidic bonds of cellulose chains randomly to produce new chain ends;

---

*Address correspondence to Sonia Couri: Rio de Janeiro Federal Institute of Science Education and Technology, Rio de Janerio, Brazil; Tel: +55 (21) 2566-7733; E-mail: sonia.couri@gmail.com

exoglucanases processively cleave cellulose chains at the ends to release soluble cellobiose or glucose; and β-glucosidases hydrolyze cellobiose to glucose [2].

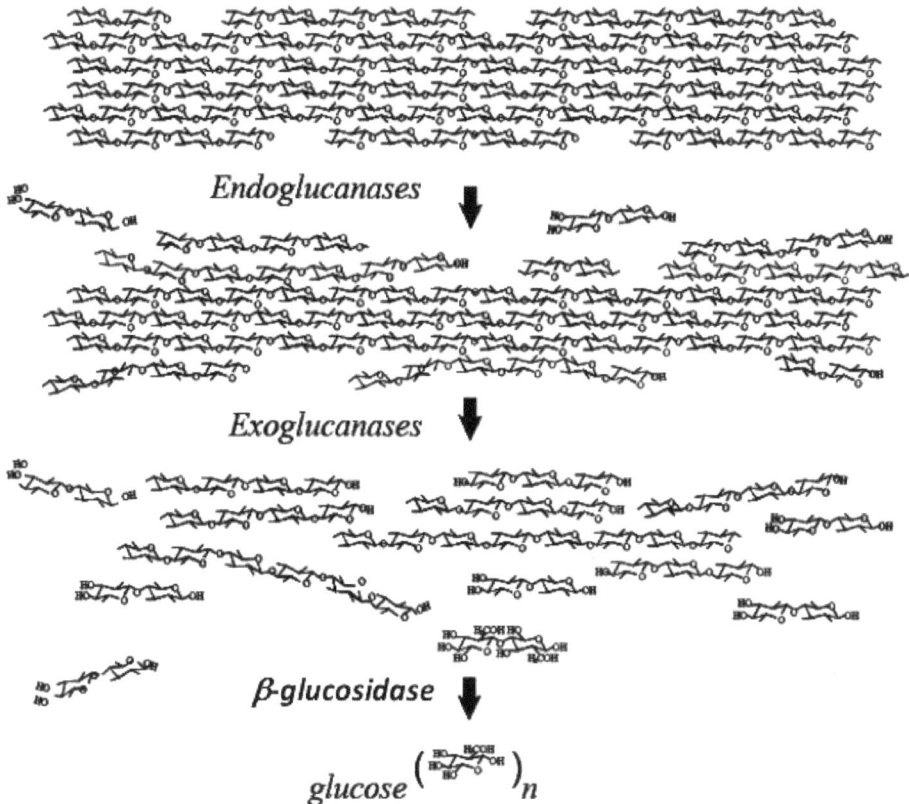

Figure 1: Synergistic action of cellulases on degrading the cellulose polymeric chain

Commercial production of cellulases began in the early 1970s, with cellulase made by the filamentous fungus *Trichoderma*. The mid-1980s saw the first large industrial uses of cellulase for stonewashing denim and as an additive for animal feeds. This was accompanied by the introduction of commercial cellulase preparations made by fungi of the genera *Aspergillus*, *Penicillium*, and *Humicola* [3].

Cellulases are the third largest industrial enzyme product worldwide, by dollar volume, but will certainly become the largest volume industrial enzyme with its application on biomass conversion into biofuels [4]. Cellulose is the most abundant natural renewable resource of the planet and energy production based on the lignocellulosic matrix is an important alternative route that has been studied and debated worldwide. However, the enzymatic hydrolysis route of cellulose conversion, although an alternative of lower environmental impact, it still requires the development of technologies that can reduce the costs of the enzymes. The costs of cellulase is considered to be one of the major constraints for the technological commercialization of the enzymatic hydrolysis of cellulose [5,6]. Besides that, cellulases have been largely applied at different industrial areas like textile, food, brewery and wine, animal feed, pulp and paper industries [7].

Cellulase preparations are expensive in the biorefinery context for two historical reasons: 1) the source of the enzymes, usually *Trichoderma reesei* was costly to grow and induce and 2) the specific performance (or activity) is low compared to other polysaccharide degrading enzymes [8]. On a protein basis, it takes from 40 to 100 times more enzyme to break down cellulose than starch [9].

The successful strategy to reduce cellulase cost must consider two key technical factors for most commercial enzyme production: enzyme specific activity and host productivity [10]. These two factors can be optimized through: 1). Microorganism selection; 2). Improvement of fermentation process conditions.

Even though a great variety of microorganisms can produce cellulases, only a few of them produce in a significant amount. Some genera of fungi that produce cellulase are *Aspergillus, Cladosporium, Fusarium, Geotrichum, Myrothecium, Paecilomyces, Penicillium,* and *Trichoderma.* Filamentous fungi are preferred over yeast and bacteria due to their higher levels of enzymes produced [11]. Besides higher productivity, fungi produce a complex mix of enzymes with high catalytic activity which are typically secreted into the growth medium [9].

The most studied and well characterized cellulase system is the one produced by the fungus *Trichoderma reesei* [12]. However, commercial preparations of cellulases made from *Trichoderma* are deficient in β-glucosidase. On the other hand, *Aspergillus niger* produces a high amount of β-glucosidase [13]. In this context, a mixed culture of *Trichoderma* and *Aspergillus* has been used for enhanced cellulase production [14].

Qualitative assays can be used in the preliminary studies for selecting cellulase producing strains. They are useful for screening large numbers of fungal and bacterial isolates for cellulase activity. A type of screening procedure for selecting cellulase producing strains is carried out on solid agar containing only crystalline cellulose as carbon source (Fig. **2**).

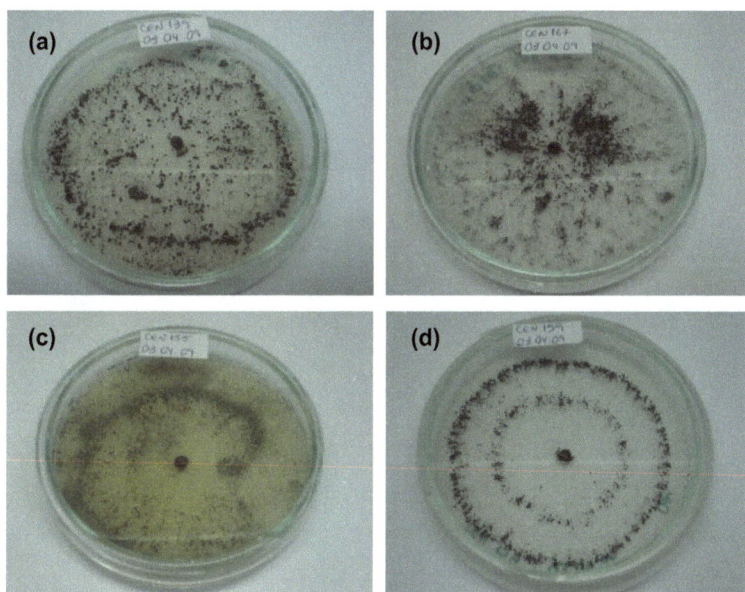

Figure 2: Screening on cellulase producing fungal strains able to grown on Avicel medium. (a): *Trichoderma* sp. CEN 139; (b) *Trichoderma* sp. CEN 167; (c) *Trichoderma* sp. CEN 155; (d) *Trichoderma* sp. CEN 159.

However, even though these tests may be convenient when having to select a large number of strains, it must be considered that they may not reflect the potential saccharification performance. Other considerations become of great importance to commercial cellulose hydrolysis, such as end-product inhibition, the addition of cellobiase activity, or reactor and process configuration. These factors will vary with different cellulase systems and significantly affect conversion efficiency [15].

Here, we describe quantitative and semi-quantitative cellulase activity assays using soluble and insoluble substrates, focusing on the need to standardized methods in order to allow a comparison among enzyme preparations.

QUANTIFYING CELLULASE ACTIVITY

There are two basic approaches to measure cellulase activity, which is measuring the individual cellulase (endoglucanase, exoglucanase, and β-glucosidase) activities, and measuring the total cellulase activity as FPAse [2]. Substrates for cellulase activity assays can be also be divided into two categories, based on their solubility in water, as soluble and insoluble substrates.

When cellulase activity is assayed on soluble substrates, it presents normal Michaelis-Menten kinetics, showing a behavior similar to other enzymes. However, when cellulases are assayed on insoluble substrates, they have very different properties, as the assays are usually nonlinear with time and with the amount of enzyme [4].

Cellulase activity assays can also be divided into three types according to: (1) the accumulation of products after hydrolysis, (2) the reduction in substrate quantity, and (3) the change in the physical properties of substrates. The majority of assays involve the accumulation of hydrolysis products, including reducing sugars, total sugars, and chromophores [2]. The most common reducing sugar assays is the dinitrosalicylic acid (DNS) method by Miller [16] which is described in detail below. Besides, glucose can be measured by an enzymatic glucose kit using coupled hexokinase and glucose-6-phosphate dehydrogenase [17], or HPLC.

Reducing Sugar Assay – DNS Method

The reducing sugar assay that has been mostly used in cellulase assays is the 3,5-dinitrosalicylic acid (DNS) method described by Miller [16] This method is based on a redox reaction between DNS and glucose (Fig. **3**), or other reducing sugars. Glucose and other sugars capable of reducing mild oxidizing agents such as ferric or cupric ion are called reducing sugars. The property of oxidizing the carbonyl carbon to a carboxyl group is the basis of qualitative tests for reducing sugar. By measuring the amount of oxidizing agent reduced by a solution of sugar, it is also possible to estimate the concentration of that sugar.

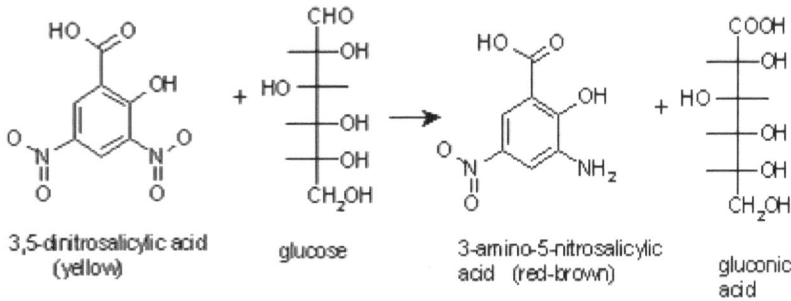

Figure 3: Reaction between 3, 5-dinitrosalicylic acid (DNS) and glucose.

The oxidation of a sugar's anomeric carbon (the reaction that defines a reducing sugar) occurs only with the linear form, which exists in equilibrium with the cyclic form(s) [18]. When the sugar anomeric carbon is involved in a glycosidic bond, that residue cannot take the linear form and therefore becomes a nonreducing sugar. In describing disaccharides or polysaccharides, the end of a chain with a free anomeric carbon (one not involved in a glycosidic bond) is commonly called the reducing end [18].

The DNS reagent contains sodium potassium tartrate (Rochelle salt), which reduces the tendency to dissolve oxygen by increasing the ion concentration in the solution. Phenol increases the amount of color produced during the color developing reaction. Sodium bisulphite stabilizes the color obtained and reacts with any oxygen present in the medium. Finally, the alkaline medium (NaOH) is required for the redox reaction between DNS and glucose, or other reducing sugars [16].

Protocol for Measuring Reducing Sugars by the DNS Method [15, 16, 19]

Materials

1. Distilled water;

2. 3,5 Dinitrosalicylic acid (DNS);

3. Sodium hydroxide;

4. Rochelle salts (sodium potassium tartrate);

5. Phenol;

6. Sodium metabisulfite.

Equipment

Spectrophotometer for measuring absorbance at 540 nm.

Procedures

Preparing the DNS Reagent

1. Dissolve 10.6 g of DNS and 19.8 g of NaOH in 1,416 mL of distilled water;

2. After complete dissolution, add 360 g of Rochelle salts (sodium potassium tartrate), 7.6 mL of melted phenol (at 50°C), and 8.3 g of sodium metabisulfite, and then mix well;

3. Titrate 3 mL of the DNS reagent using 0.1 M HCl using the phenolphthalein end point pH check. It should take 5–6 mL of HCl for a transition from red to colorless. Add NaOH if required (2 g of NaOH added = 1 mL of 0.1 M HCl used for 3 mL of the DNS reagent).

Measuring Reducing Sugars

1. Place 1-2 mL sample in a test tube and add 3 mL DNS reagent;

2. Boil all tubes for 5.0 min in a vigorously boiling water bath containing sufficient water to cover the portions of the tubes occupied by the reaction mixture plus reagent. All samples, controls, blanks, and glucose standards should be boiled together;

3. After boiling, transfer to a cold ice water bath;

4. Dilute all tubes (samples, blanks, standards and controls). Add 0.200 mL of color developed reaction mixture plus 2.5 mL of water in a spectrophotometer cuvette and mixture well;

5. Measure absorbance at 540 nm;

6. Draw a standard sugar curve (sugar along the x-axis *vs.* absorbance at 540 nm along the y-axis) following the volumes and standard

concentrations particular of each enzyme assay. Different color intensities are generated by the different glucose concentrations, as shown in Fig. **4**;

Figure 4: Glucose standard showing the variation of colors generated by the different concentrations of glucose.

7. The data for the standard curve should closely fit a calculated straight line, with the correlation coefficient for this straight line fit being near to one, as shown in Fig. **5**;

8. Use this standard curve to determine the amount of glucose released for each sample tube after subtraction of enzyme blank.

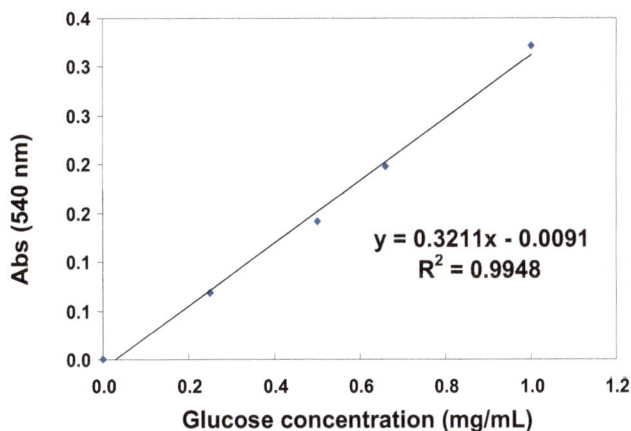

$$y = 0.3211x - 0.0091$$
$$R^2 = 0.9948$$

Figure 5: Glucose standard curve indicating the angular coefficient of the standard linear regression.

Modified DNS Reagent

The use of a modified DNS reagent has being adopted in some laboratories in order to avoid the use of phenol due to its toxicity. Since phenol is responsible to increase the amount of color produced during the color developing reaction, the values of absorbance obtained using this modified reagent is lower than the ones described before. Consequently, the slope of the curve will be smaller, as can be seen in Fig. **6**.

Preparing the Modified DNS Reagent

1. Dissolve 300 g of sodium potassium tartarate tetrahydrate and 16 g of sodium hydroxide in 500 mL of distilled water;

2. When the solution is clear, add 10 g of DNS;

3. Bring the volume to 1000 mL with distilled water in a volumetric flask;

4. Store the solution in amber bottle (or dark bottle) protected from light;

5. Follow the same procedure described above for measuring reducing sugars.

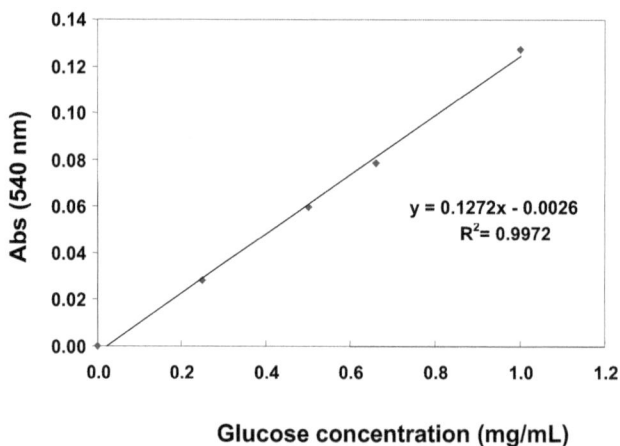

$y = 0.1272x - 0.0026$
$R^2 = 0.9972$

Figure 6: Glucose standard curve using the modified DNS reagent indicating the angular coefficient of the standard linear regression.

FILTER PAPER ASSAY (FPA)

Filter paper assay (FPA) is the most common total cellulase activity assay recommended by the International Union of Pure and Applied

Chemistry (IUPAC), which procedure is described by Ghose [15]. This assay was developed by Mandels *et al.* [20] to measure the hydrolysis activity of cellulases using filter paper as substrate, a readily available and reproducible substrate. Factors that affect sensitivity and reproducibility often result from the fact that most natural cellulase complexes tend to have a deficiency of β-glucosidase activity. The level of β-glucosidase in an enzyme preparation may affect the result of cellulase assays, in particular the FPA [21].

Two suggestions can be found in the literature to improve the reproducibility of the Mandels' filter paper assay: add supplemental β-glucosidase and increase the boiling time for color development. Coward-Kelly *et al.* [22] provided data that supports adding supplemental β-glucosidase, having an increased assay response by 56%. In terms of boiling time, they showed that here is no need for additional boiling time, being 5 minutes sufficient. For maximum reproducibility, it is essential that the water bath vigorously boil so that temperature excursions are minimized [22].

Protocol for FPA Assay [15, 19, 23]

Materials

1. Citrate buffer (0.050 mol/L, pH 4.8);

2. Whatman No. 1 filter paper strip (1.0 x 6.0 cm);

3. 13 x 100 test tube;

4. DNS reagent.

Equipment

1. Water bath capable of maintaining a temperature of 50°C ± 0.1 °C;

2. Spectrophotometer suitable for measuring absorbance at 540 nm.

Procedure (Fig. 7)

1. Place a rolled filter paper strip into a 13 x 100 test tube;

2. Add 1.0 mL of 0.050 mol/L citrate buffer, pH 4.8 to the tube; the buffer should saturate the filter paper strip;

3. Equilibrate tubes with buffer and substrate to 50°C;

4. Add 0.5 mL of enzyme diluted appropriately in citrate buffer. At least two dilutions must be made of each enzyme sample, with one dilution releasing slightly more than 2.0 mg of glucose (absolute amount) and one slightly less than 2.0 mg of glucose;

5. Incubate at 50°C for 60 min;

6. At the end of the incubation period, remove each assay tube from the 50°C bath and stop the enzyme reaction by immediately adding 3.0 mL DNS reagent and mixing;

7. Boil all tubes for exactly 5.0 minutes in a vigorously boiling water bath. All samples, controls, blanks, and glucose standards should be boiled together. After boiling, transfer to a cold ice water bath;

8. Let the tubes sit until all the pulp has settled, or centrifuge briefly;

9. Dilute all tubes (assays, blanks, standards and controls) in water. Add 0.200 mL of color developed reaction mixture plus 2.5 mL of water in a spectrophotometer cuvette and mixture well;

10. Determine color formation by measuring absorbance against the reagent blank at 540 nm.

Blank and Controls

1. Reagent blank: 1.5 mL citrate buffer;

2. Enzyme control: 1.0 mL citrate buffer + 0.5 mL enzyme dilution (prepare a separate control for each dilution tested);

3. Substrate control: 1.5 mL citrate buffer + filter-paper strip;

4. Blanks, controls and glucose standards should be incubated at 50°C along with the enzyme assay tubes, and then "stopped" at the end of 60 minutes by addition of 3.0 mL of DNS reagent.

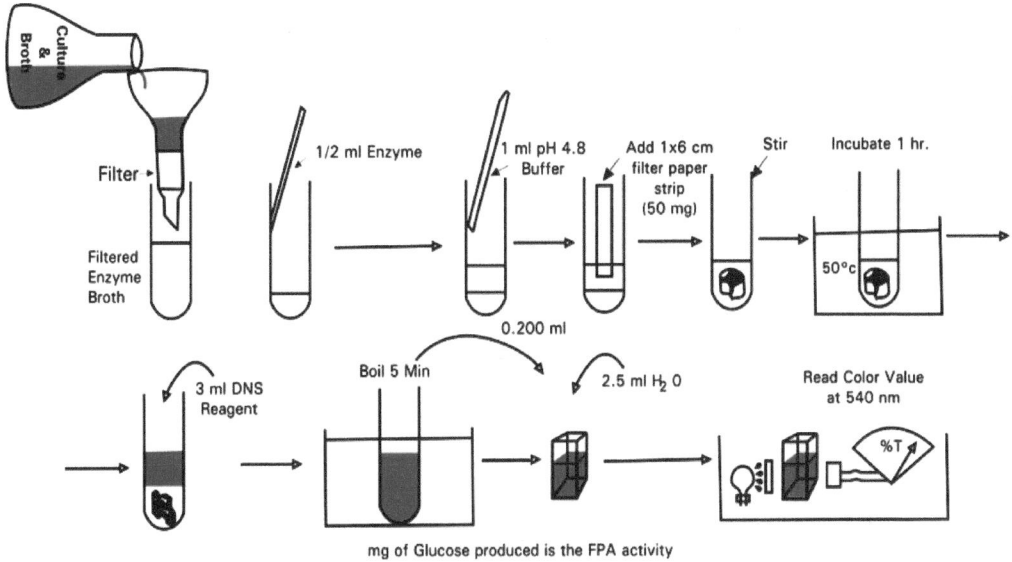

Figure 7: Filter paper assay procedure (Adapted from [24]).

Glucose Standard Curve

1. A working stock solution of anhydrous glucose (10 mg/mL) should be prepared;

2. Dilutions are made from the working stock as follows:

1.0 mL glucose standard + 0.5 mL buffer = 1:1.5 = 6.7 mg/mL (3.35 mg/0.5 mL).

1.0 mL glucose standard + 1.0 mL buffer = 1:2 = 5 mg/mL (2.5 mg/0.5 mL).

1.0 mL glucose standard + 2.0 mL buffer = 1:3 = 3.3 mg/mL (1.65 mg/0.5 mL).

1.0 mL glucose standard + 4.0 mL buffer = 1:5 = 2 mg/mL (1.0 mg/0.5 mL).

3. Glucose standard tubes should be prepared by adding 0.5 mL of each of the above glucose dilutions to 1.0 mL of citrate buffer in a test tube.

Calculation

1. Calculate the delta absorbance of dilute enzyme solutions for each sample by subtraction of the absorbance of enzyme control;

2. Calculate the real glucose concentrations released by each sample according to a standard sugar curve;

3. Estimate the concentration of enzyme which would have released exactly 2.0 mg of glucose by means of a plot of glucose liberated against the logarithm of enzyme concentration. To find the required enzyme concentration take two data points that are very close to 2.0 mg and draw a straight line between them, use this line to interpolate between the two points to find the enzyme dilution that would produce exactly 2.0 mg glucose equivalents of reducing sugar;

4. Calculate the FPA of the original concentrated enzyme solution in terms of FPU/mL:

$$\text{Filter Paper Activity} = \frac{0.37 \text{ units/mL}}{[\text{enzyme}] \text{ releasing } 2.0 \text{ mg of glucose}}$$

Where:

0.37 μmol/min/mL represents 2 mg glucose/(0.18 mg/μmol) × 0.5 mL × 60 min;
[enzyme] represents the proportion of original enzyme solution present in the directly tested enzyme dilution (that dilution of which 0.5 mL is added to the assay mixture).
For example a 1:20 dilution of the working stock of enzyme will have "concentration" of 0.05.

FPAse Production and Quantification

The enzymatic complex produced by a selected strain of the filamentous fungus *Aspergillus fumigatus* cultivated using different agro-industrial residues as substrate in solid state fermentation (SSF) was analyzed in terms of total cellulase. Fig. **8** shows the effect of carbon source on the kinetics of FPAse. Higher FPAse were obtained when wheat bran as well as a mixture of sugarcane bagasse and wheat bran (1:1) were used as solid substrates, reaching values of up to 5.0 FPU/g of solid substrate. This result is rather interesting when compared to others substrates evaluated, since soybean bran resulted in FPAse activity values that were less than half of those obtained with wheat bran (2.4 IU/g). The other substrates evaluated did not result in significant FPAse activities. The maximum activity of FPAse occurred after 96 h of incubation.

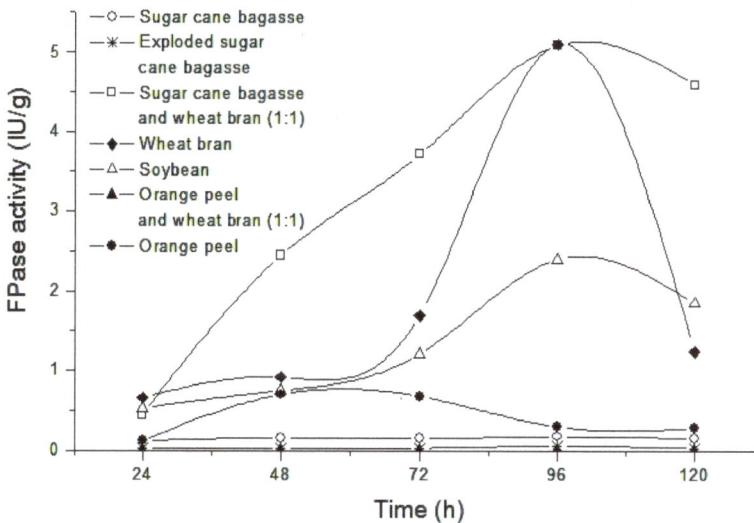

Figure 8: Effect of different carbon sources on the production of FPAse by *A. fumigatus* under solid-state fermentation (SSF).

Alternative Methods for Measuring FPA Activity

Some available literature describes alternative methods for measuring FPAse activity stating that the IUPAC filter paper assay is time-consuming, labor-intensive, and requires large amounts of reagents [25, 26].

In order to circumvent these issues, Xiao *et al.* [25] developed a microplate-based method to determine total cellulase activity when having a large sample numbers. To achieve this, the enzymatic reaction volume was reduced from 1.5 mL to 60 μl. The modified 60 μl format FPA was carried out in 96-well assay plates and statistical analyses showed that the cellulase activities of commercial cellulases from *Trichoderma reesei* and *Aspergillus* species determined in the 60 µl format were not significantly different from the activities measured with the standard FPA. According to the authors, this new assay can be used to rapidly and easily assay a large number of samples, reducing the amount of chemicals and enzyme used by 25-fold relative to the standard IUPAC method.

Decker *et al.* [26] report an automated FPA method based on a Cyberlabs C400 robotics deck equipped with customized incubation, reagent storage, and plate-reading capabilities that allow rapid evaluation of cellulases acting on cellulose.

ENDOGLUCANASE ASSAYS

Endoglucanases (EC 3.2.1.4) or 1,4-β-D-glucan-4-glucanohydrolases are enzymes of the cellulase complex responsible for initiating the hydrolysis process. Such enzymes hydrolyze randomLy the inner regions of the amorphous structure of the cellulose fiber, releasing oligosaccharides from different degrees of polymerization and, consequently, new terminals, including reducing and non-reducing ends [27]. Endoglucanase activity assays are often performed using as substrate the ionic substituted carboxymethyl cellulose (CMC). That is because endoglucanases, also called CMCase, cleave intramolecular β-1,4-glucosidic bonds randomLy, resulting in a dramatic reduction in the degree of polymerization (*i.e.*, specific viscosity) of CMC [2].

Endoglucanase activity can be measured based on a reduction in substrate viscosity and/or an increase in reducing ends determined by a reducing sugar assay, such as the assay based on the Ghose methodology [15]. There are also semi-quantitative methods for endoglucanase activities based on staining residual polysaccharides (CMC, cellulose) on agar plates with various dyes which are adsorbed by long chains of polysaccharides. After hydrolysis, staining, and washing, halo zones can be observed in the colored background [2]. Zymogram is also a useful tool to estimate endoglucanase profiles produced by microorganisms. This technique provides the visualization of the enzymatic activity, due to the utilization of the substrate (CMC) copolymerized with the polyacrylamide gel. Although there are some other methods to measure endoglucanase activity, the protocol described here is the most common one described in literature [15, 19].

Protocols for Endoglucanase Assays

Endoglucanase Activity [15, 19]

Materials

1. Citrate buffer (0.050 mol/L, pH 4.8);

2. CMC (2% w/v) in citrate buffer;

3. DNS reagent.

Equipment

1. Water bath capable of maintaining a temperature of 50°C ± 0.1 °C;

2. Spectrophotometer suitable for measuring absorbance at 540 nm.

Procedure

1. Add 0.5 mL of enzyme diluted appropriately in citrate buffer to a test tube. At least two dilutions should be made of each enzyme sample analyzed. One dilution should release slightly more and one slightly less than 0.5 mg (absolute amount) of glucose in the reaction conditions;

2. Equilibrate the enzyme solution and substrate solution at 50°C;

3. Add 0.5 mL of substrate solution, mix well and incubate at 50°C for 30 min;

4. Add 3.0 mL DNS and mix well;

5. Boil for 5.0 min in a vigorously boiling water bath containing sufficient water. All samples, blanks and glucose standards should be boiled together. After boiling, transfer immediately to a cold water bath;

6. Add 0.200 mL of color developed reaction mixture plus 2.5 mL of water in a spectrophotometer cuvette and mixture well;

7. Measure the color formed at 540 nm. The color formed in the enzyme blank should be subtracted from that of the sample tube;

8. Convert the absorbance of the sample into glucose production during the reaction using a glucose standard curve (DNS method).

Blank and Controls

Espectro Zero: 0.5 mL substrate solution + 0.5 mL citrate buffer

Enzyme control: 0.5 mL substrate solution + 0.5 mL enzyme dilution

Glucose standard:

A working stock solution of anhydrous glucose (2.0 mg/mL) should be prepared. Dilutions are made from the working stock as follows:

0.125 mL glucose standard + 0.875 mL buffer = 0.25 mg/mL

0.250 mL glucose standard + 0.750 mL buffer = 0.50 mg/mL

0.330 mL glucose standard + 0.670 mL buffer = 0.66 mg/mL

0.500 mL glucose standard + 0.500 mL buffer = 1.0 mg/mL

Calculation

1. Construct a linear glucose standard curve using the absolute amounts of glucose (mg/0.5 mL) plotted against absorbance at 540 nm;

2. Using this standard curve, convert the absorbance values of the sample tubes (after subtraction of enzyme blank) into glucose (= mg of glucose produced during the reaction);

3. Convert the dilutions used into enzyme concentrations:

$$\text{Concentration} = \frac{1}{\text{Dilution}} \qquad \frac{(= \text{volume of enzyme in dilution})}{\text{total volume of dilution}}$$

4. Estimate the concentration of enzyme which would have released exactly 0.5 mg of glucose by plotting glucose liberated (2) against enzyme concentration (3) on semi logarithmic graph paper;

5. Calculate the CMCase of the original concentrated enzyme solution in terms of IU/mL:

$$\text{CMCase} = \frac{0.185}{[\text{enzyme}] \text{ releasing } 0.5 \text{ mg of glucose}} \text{ units/mL}$$

Where 0.185 µmol/min/mL represents 0.5 mg glucose/(0.18 mg/µmol) × 0.5 mL × 30 min. Enzyme concentration represents the proportion of original enzyme solution present in the directly tested enzyme dilution (that dilution of which 0.5 mL is added to the assay mixture). For example a 1:20 dilution of the working stock of enzyme will have "concentration" of 0.05.

Alternative Procedures

There are often variations in the endoglucanase activity assay procedure in terms of the concentration of substrate and time of incubation (substrate-enzyme reaction). A procedure that is being used in some laboratories adopted a substrate solution of 4% CMC in a 10 min reaction time. Even thought, the experimental sequence steps are the same as described before, care should be taken during calculation of enzymatic activity in order to take these variations into account. When these alternative procedures are used the following calculation should be employed.

Calculation

1. Construct a glucose standard curve using the glucose concentration plotted against absorbance at 540 nm;

2. Using this standard curve, convert the absorbance values of the sample tubes (after subtraction of enzyme blank) into glucose concentration;

3. Calculate the CMCase of the original concentrated enzyme solution in terms of IU/mL:

$$\text{CMCase (IU/mL)} = \frac{\text{dilution*Glucose Concentration (mg/mL)*Total reaction volume (mL)}}{0.18 * \text{reaction time (min)} * \text{Total volume of enzyme dilution (mL)}}$$

Where
0.18 represents the conversion of 1 mg to µmol of glucose.

Zymogram

Material

1. Citrate buffer (0.050 mol/L, pH 5.0);

2. NaCl (1 M) solution;

3. 10% (v/v) acetic acid solution;

4. 0.2% Congo Red solution;

5. Polyacrylamide gel electrophoresis reagents.

Equipment

1. Electrophoresis system;

2. Transilluminator with visible light.

Procedure

Crude enzyme preparations should be first fractionated by native polyacrylamide gel electrophoresis (PAGE) using 10% acrylamide gel containing CMC 0.5% (w/v). For developing endoglucanase activity, the PAGE gel should be incubated for 15 min in 0.05 mol/L sodium citrate buffer (pH 5.0). After removal of the buffer, the gels incorporating CMC substrate (0.5%, w/v) should be stained with 0.2% (w/v) Congo Red for 15–30 min and then destained with 1 mol/L NaCl until bands appeared. Gels are then counterstained with 10% (v/v) acetic acid. Bands corresponding to endoglucanase appear as clear zones against a dark background after distaining with 1 mol/L NaCl followed by treatment with acetic acid solution (Fig. **9**).

Congo Red Test on Agar Medium [19]

Material

1. Congo red solution (1 g/l) prepared by dissolving 100 mg of Congo red (Fig. **10**) in 99 mL of water and 1% ethanol;

2. NaCl (1 mol/L) solution;

3. Sodium phosphate buffer (0.1 mol/L, pH 6.5);

4. CMC (1% w/v, low viscosity) in 1.5% agar medium. Dissolve CMC before adding agar and autoclave.

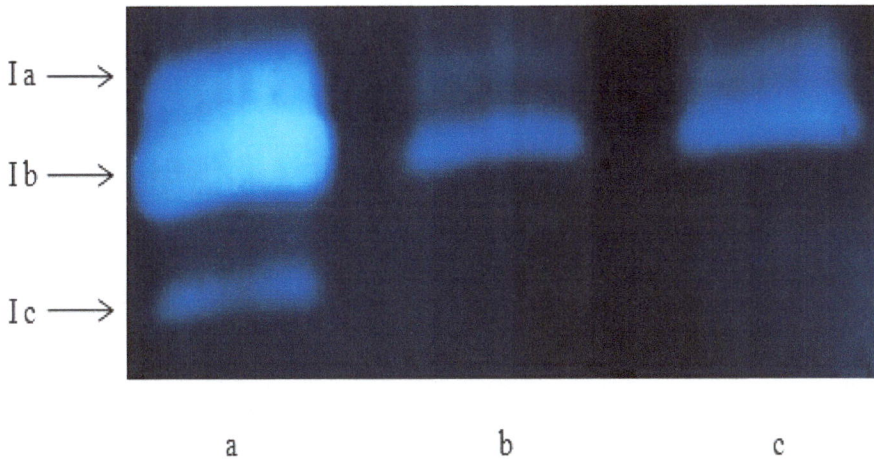

Figure 9: Zymogram showing endoglucanases produced by *A. fumigatus* using as substrate (a) soybean, (b) orange peel and wheat bran, (c) cane bagasse and wheat bran.

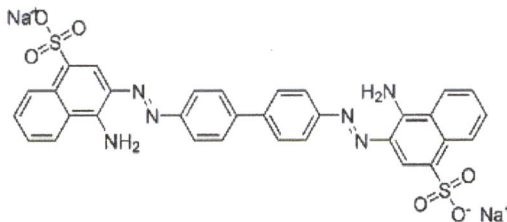

Figure 10: Congo red structure.

Procedure

1. Inoculate the endoglucanase-secreted microorganisms on the solid CMC medium. The growth time depends on the growth rate of the microorganism and enzyme activity;

2. Stain a 9-cm Petri dish by adding 20 mL of Congo red solution at room temperature for 30 min;

3. Rinse the residual dye on the dish using distilled water;

4. Destain Congo red with ~20 mL of 1 mol/L NaCl solution for 30 min. If the halos are not clear, destain the dish by another ~20 mL of NaCl solution;

5. Detect the clear, weak yellow halos for endoglucanase activity with the red background. In strains producing cellulases there is a halo of light color with orange edges, indicative of areas of hydrolysis, as shown in Fig. **11**;

6. This halo is measured for subsequent calculation of the enzymatic index (EI):

Enzymatic index (EI) = $\dfrac{\text{Diameter of the halo of hydrolysis}}{\text{Diameter of the colony}}$

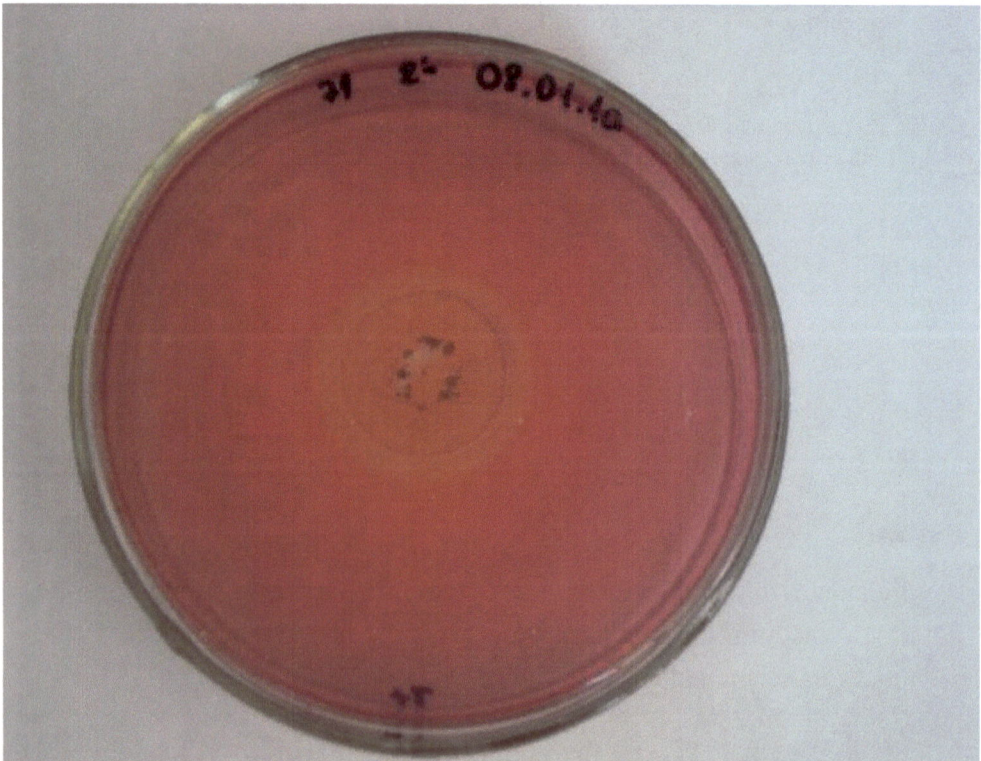

Figure 11: Congo red test for *Thricoderma harzianum*.

Endoglucanase Production and Quantification

The potential of different strains of the filamentous fungi *Trichoderma* in producing endogucanases was evaluated by growing the microorganisms in test tubes containing a nutrient medium and a strip of filter paper as the sole carbon source. As shown in Fig. **12**, the system simulates the process of submerged

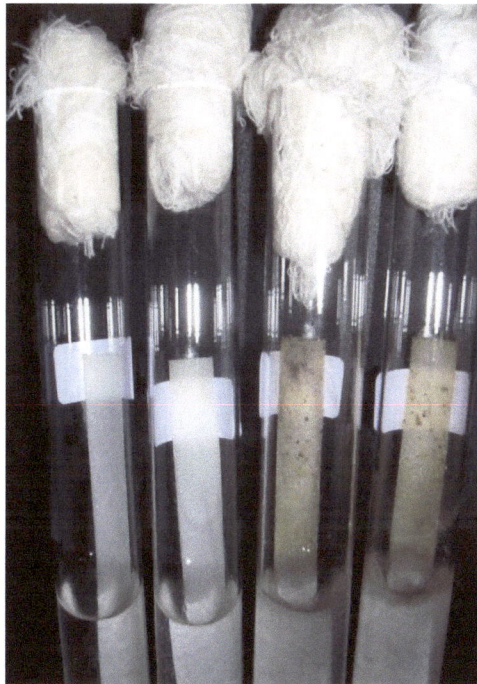

Figure 12: Production of endoglucanases by *Trichoderma* strains on fermentation tubes.

fermentation with the difference of using an insoluble substrate. The microorganisms grow preferentially in the solid phase in order to have access to the carbon source, capturing the other nutrients by diffusion. In the experiment herein, the tubes were kept incubated during four weeks and every week one tube was removed for enzymatic analysis. The results for endoglucanase activity are shown in Fig. **13**.

It can be observed that *Trichoderma harzianum* CEN 139, *T. harzianum* CEN 155 and *T. sp* CG 104 NH were the strains that showed higher endoglucanase activities when compared to the other strains.

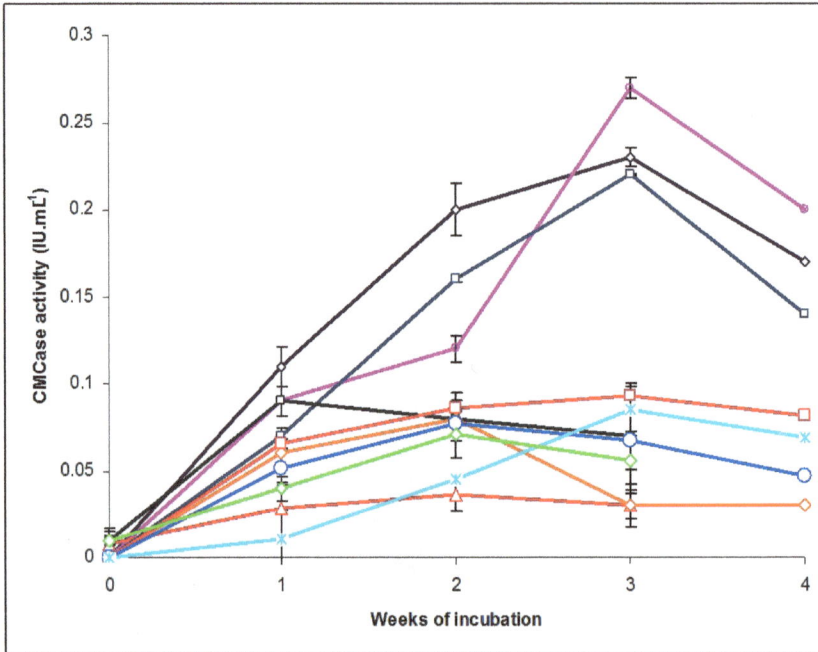

Figure 13: Endoglucanase production by different *Trichoderma* strains: (-O-)*T. harzianum* CEN 139; (-□-) *T. harzianum* CEN 155; (-○-)*T. sp* 104 NH; (-□-) *T. asperelum* CEN 201; (-*-) *T. harzianum* CEN 241; (-◇-) *T. harzianum* CEN 248; (-O-) *T. harzianum* CEN 238; (-□-) *T. sp* CEN 156; (-△-) *T. koningii* CEN 142; (-◇-) *T. sp* CEN 159.

Exoglucanases

Exoglucanase enzymes cleave the accessible ends of cellulose molecules to liberate glucose and cellobiose [2]. Exoglucanases are composed by cellobiohydrolases (CBHs) and glucanohydrolases (GHs). The GHs (EC 3.2.1.74) are highly important on hydrolysis of cellulose fiber because they are able to directly release glucose [27]. The CBHs (EC 3.2.1.91) can be further divided into two types: type I enzymes (CBHs I), which hydrolyze reducing terminals, whereas type II (CBHs II) hydrolyze non-reducing terminals. The activity of purified exoglucanases is often estimated using Avicel as substrate because of its highest ratio of end/accessibility among insoluble cellulosic substrates [2].

Protocol for Exoglucanase Assay [19]

Materials

1. Avicel (FMC PH 101 or PH 105 or Sigmacell 20);

2. Sodium acetate buffer (0.1 mol/L, pH 4.8);

3. DNS reagent.

Equipment

1. Water bath capable of maintaining a temperature of 50°C ± 0.1 °C;

2. Spectrophotometer suitable for measuring absorbance at 540 nm.

Procedure

1. Suspend 1.25% (w/v) Avicel in acetate buffer;

2. Add 1.6 mL of Avicel suspension solution into the tubes;

3. Dilute a series of enzyme solutions with acetate buffer;

4. Equilibrate the substrate and enzyme solutions at 50°C;

5. Add 0.4 mL of the dilute enzyme solutions to the Avicel substrate and mix well;

6. Incubate at 50°C for 2 h;

7. Stop the reaction by submerging the tubes in ice-cooled water bath;

8. Withdrew 1 mL of hydrolysate into microcentrifuge tubes and centrifuge the sample at 13,000 g for 3 min;

9. Prepare enzyme blanks (0.4 mL of diluted enzymes and 1.6 mL of acetate buffer) and substrate blank (0.4 mL of acetate buffer and 1.6 mL of 1.25% (w/v) Avicel suspension buffer);

10. Determine the total soluble sugars in the supernatant by the DNS method;

11. Calculate the enzyme activity on the basis of a linear relationship between the total soluble sugar release and enzyme dilution. One unit of

exoglucanase activity is defined as the amount of enzyme that releases one micromole of glucose equivalent per minute from Avicel, as follows:

Exoglucanase (IU/mL) = $\dfrac{\text{dilution*Glucose Concentration (mg/mL)*Total reaction volume(mL)}}{0.18 * \text{reaction time (min) * Total volume of enzyme dilution (mL)}}$

Where 0.18 represents the conversion of 1 mg to μmol of glucose.

β-glucosidase

The third and last major group of enzymes of the cellulolytic complex includes the enzymes β-glucosidase or β-glycoside glucohydrolase (EC 3.2.1.21), its systematic name. β-Glucosidase can cleave b-1,4-glucosidic bonds of soluble substrates, including cellobiose, longer cellodextrins with a degree of polymerization from 3 to 6, and chromogenic substrates, such as p-nitrophenyl- β-D-1,4-glucopiranoside [2].

Protocols for β-Glucosidase Assay

Following protocols for β-glucosidase assays include determination of activity using cellobiose and p-nitrophenyl- β-D-1,4-glucopiranoside as substrate (Fig. **14**), as well as a zymogram assay.

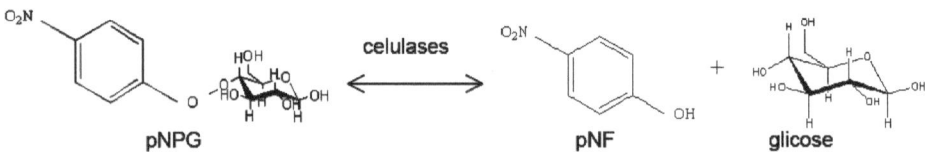

Figure 14: Reaction between pNPG and cellulases.

β-glucosidase Activity Using Cellobiose [15,19]

Materials

1. Citrate buffer (0.050 mol/L, pH 4.8);

2. 15.0 mmol/L Cellobiose in citrate buffer (freshly made substrate solution).

Equipment

1. Water bath capable of maintaining a temperature of 50°C ± 0.1 °C.;

2. Spectrophotometer for measuring absorbance at 540 nm;

Procedure

1. Add 1.0 mL of enzyme, diluted in citrate buffer, to a small test tube. At least two dilutions must be made of each enzyme sample. One dilution should release slightly more and one slightly less than 1.0 mg (absolute amount) of glucose in the reaction conditions;

2. Equilibrate at 50°C;

3. Add 1.0 mL substrate solution and mix well;

4. Incubate at 50°C for 30 min;

5. Terminate the reaction by immersing the tube in boiling water for 5.0 min;

6. Transfer the tube to a cold water bath;

7. Prepare substrate blank (1.0 mL of cellobiose solution and 1.0 mL of citrate buffer) and enzyme blanks (1.0 mL of citrate buffer and 1.0 mL of diluted enzyme solution). Treat substrate and enzyme blanks identically as the experimental tubes (*i.e.*, equilibrate at 50°C, heat, boil, and cool);

8. Determine glucose produced using a standard procedure (commercial kit based on the glucose oxidase (GOD) reaction);

9. Measure the absorbance of all solutions based on the substrate blank.

Calculation

1. Calculate the delta absorbance of dilute enzyme solutions by subtracting absorbance of the respective enzyme blanks. Calculate the

real glucose concentrations released according to a standard glucose curve by the enzyme kit;

2. Multiply by 2 to convert glucose concentration into absolute amounts (mg);

3. Convert the dilutions used into enzyme concentrations:

$$\text{Concentration} = \frac{1}{\text{Dilution}} \qquad \left(= \frac{\text{volume of enzyme in dilution}}{\text{total volume of dilution}} \right)$$

-   Estimate the concentration of enzyme which would have released exactly 1.0 mg of glucose by plotting glucose liberated (2) against enzyme concentration (3) on semi logarithmic graph paper.

-   Calculate cellobiase activity (IU/mL):

$$\text{Cellobiase} = \frac{0.0926}{[\text{Enzyme}] \text{ releasing } 1.0 \text{ mg of glucose}} \quad \text{units/mL}$$

Where 0.0926 μmol/min/mL represents 1.0 mg glucose/(0.18 mg/μmol) × 2.0 mL × 30 min. Enzyme concentration represents the proportion of original enzyme solution present in the directly tested enzyme dilution (that dilution of which 1.0 mL is added to the assay mixture). For example a 1:20 dilution of the working stock of enzyme will have "concentration" of 0.05.

β-glucosidase Activity Using pNPG [19]

Materials

1. Sodium acetate buffer, (0.1 mol/L, pH 4.8);

2. pNPG (5 mmol/L) in acetate buffer;

3. Glycine buffer (0.4 mol/L, pH 10.8);

4. p-Nitrophenol (pNP; 20 g/l) in acetate buffer.

Equipment

1. Water bath capable of maintaining a temperature of 50°C ± 0.1 °C;

2. Spectrophotometer for measuring absorbance at 540 nm.

Procedure

1. Add 1.0 mL of pNPG solution and 1.8 mL of acetate buffer into test tubes;

2. Equilibrate at 50°C;

3. Prepare the enzyme dilution series;

4. Add 0.2 mL of diluted enzymes into the tubes containing the substrate, and mix well;

5. Enzyme blanks: Add 0.2 mL of diluted enzymes into the tubes containing 2.8 mL of acetate buffer, and mix well; Substrate blank: Add 1.0 mL of pNPG solution and 2.0 mL of acetate buffer into test tubes;

6. Incubate all tubes at 50°C for 15 or 30 min;

7. Add 4.00 mL of glycine buffer to stop the reaction;

8. Measure the absorbance of liberated products of p-nitrophenol at 430 nm based on the substrate blank;

9. Read the net absorbance of the enzyme solutions by subtracting readings of the enzyme blanks;

10. Determine p-nitrophenol release on the basis of the known concentration of p-nitrophenol diluted by glycine at 430 nm;

11. Calculate the enzyme activity on the basis of the linear range between absorbance and enzyme concentrations.

Zymogram

Material

1. Citrate buffer (0.050 mol/L, pH 5.0);

2. NaCl (1 mol/L) solution;

3. 10% (v/v) acetic acid solution.

Equipment

1. Electrophoresis system;

2. Transilluminator with visible light.

Procedure

β-Glucosidase can be located after nondenaturing polyacrylamide gel electrophoresis by incubating the gel with 0.1% esculin and 0.03% ferric chloride. The esculetin released from esculin by β-glucosidase action reacts with ferric ion to produce a black band, corresponding to the β-glucosidase, against the transparent background [28], as shown in Fig. **15**.

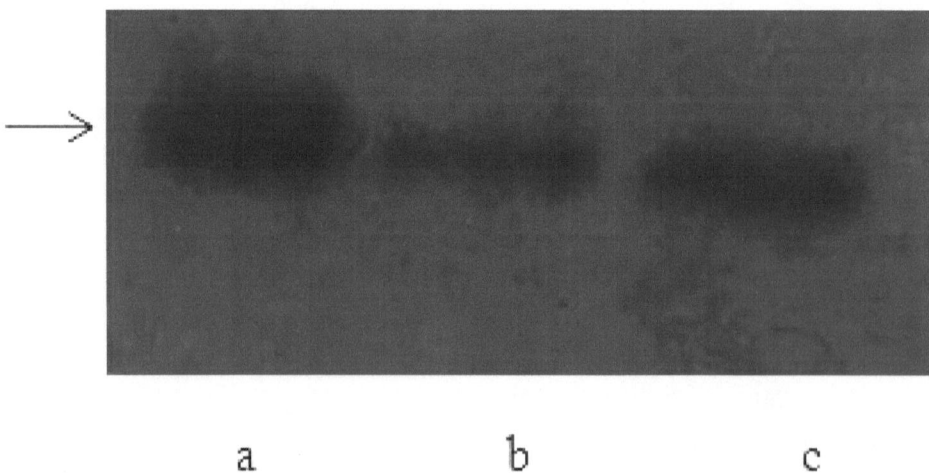

Figure 15: Zymogram showing β-glucosidase in (a) wheat bran, (b) cane bagasse and wheat bran, c) soybean.

β-Glucosidase Production and Quantification

The enzymatic complex produced by a selected strain of the filamentous fungus *Aspergillus fumigatus* cultivated using different agro-industrial residues as substrate in solid state fermentation (SSF) was analyzed in terms of β-glucosidase (Fig. **16**). Wheat bran also resulted in maximal production of β-glucosidase (105.82 IU/g), followed by a mixture of sugarcane bagasse and wheat bran (1:1) (38.04 IU/g). The highest activity occurred after 96 h of incubation.

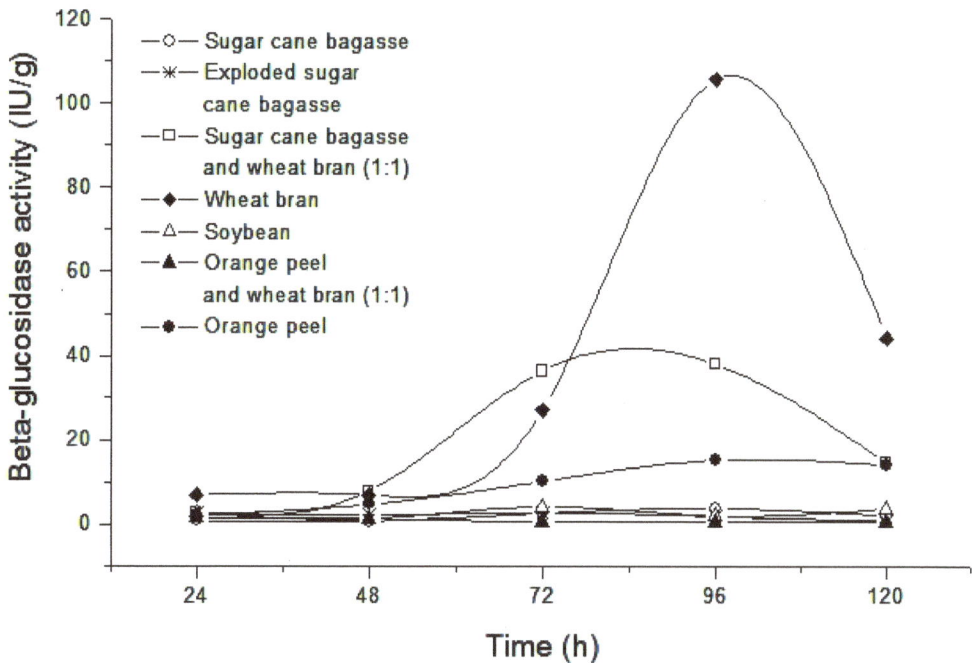

Figure 16: Effect of different carbon sources on the production of β-glucosidase by *A. fumigatus* under solid-state fermentation (SSF).

ACKNOWLEDGEMENTS

The authors thank Embrapa, MCT-CNPq, and Fapesp (all from Brazil) for the financial support.

CONFLICT OF INTEREST

The authors confirm that this chapter content has no conflict of interest.

REFERENCES

[1] WILSON, D. B,;·IRWIN D. C. Genetics and Properties of Cellulases. Advances in Biochem Eng Biotechnol 1999, 65: 1-21.

[2] ZHANG, Y. H. P.; HIMMEL, M.; MIELENZ, J. R. Outlook for cellulase improvement: screening and selection strategies. Biotechnol Adv 2006, 24(5):452–481.

[3] TOLAN, J. S.; FOODY, B. Cellulase from submerged fermentation. Advances in Biochem Eng Biotechnol 1999, 65: 41-67.

[4] WILSON, D. B. Aerobic Microbial Cellulase Systems. In: Biomass Recalcitrance – Deconstructing the Plant Cell Wall for Bioenergy, Blackwell Publishing, 2008.

[5] WALKER, L. P.; WILSON, D. B. Enzymatic hydrolysis of cellulose: An Overview. Bioresouce Technol 1991, 36: 3-14.

[6] EVELEIGH, D. E. Cellulase: a perspective. Phil Trans R Soc Lond Ser A 1987, 321: 435-47.

[7] SUKUMARAN, R. K.; SINGHANIA, R. R.; PANDEY, A. Microbial cellulases – Production, applications and challenges. J Sci Ind Res India 2005, 64: 832-844.

[8] HIMMEL, M. E.; PICATAGGIO, S. K. Our Challenge Is to Acquire Deeper Understanding of Biomass Recalcitrance and Conversion. In: Biomass Recalcitrance – Deconstructing the Plant Cell Wall for Bioenergy, Blackwell Publishing, 2008.

[9] MERINO, S. T.; CHERRY, J. Progress and challenges in enzyme development for biomass utilization. Adv Biochem Engin/Biotechnol 2007, 108: 95-120.

[10] SHEEHAN, J.; HIMMEL, M. Enzymes, Energy, and the Environment: A Strategic Perspective on the U.S. Department of Energy's Research and Development Activities for Bioethanol. Biotechnol Prog 1999, 15: 817-827.

[11] BAKRI, Y. P.; JACQUES, P.; THONART, P. Xylanase production by Penicillium canescens 10-10c in solid-state fermentation. Appl Biochem Biotechnol 2003, 108: 737–48.

[12] KADAN, K. L. Cellulase Production. In: Wyman CE Eds. Handbook on bioethanol: production and utilization. Taylor and Francis, Washington, DC, 1996.

[13] RASHID, M. H.; RAJOKA, M. I.; SIDDIQUI. K. S.; SHAKOORI, A. R. Kinetic properties of chemically modified b-glucosidase from *Aspergillus niger* 280, Pak J Zool 1997, 29: 354–63.

[14] AHAMED, A. Vermette P. Enhanced enzyme production from mixed cultures of *Trichoderma reesei* RUT-C30 and *Aspergillus niger* LMA grown as fed batch in a stirred tank bioreactor. *Biochem Eng J* 2008; 42: 41–46.

[15] GHOSE, T. K. Measurement of cellulase activities. Pure Appl Chem 1987, 59: 257-268.

[16] MILLER, G. L. Use of dinitrosalicylic acid reagent for determination of reducing sugar. Anal Chem 1959; 31(3): 426–28.

[17] ZHANG,Y. H. P; LYND, L. R. Kinetics and relative importance of phosphorolytic and hydrolytic cleavage of cellodextrins and cellobiose in cell extracts of *Clostridium thermocellum*. Appl Environ Microbiol 2004, 70:1563–9.

[18] NELSON, D. L.; COX, M. M. Lehninger Principles of Biochemistry. 4ed. Nova Iorque: W. H. Freeman, 2004.

[19] ZHANG, Y. H. P.; HONG, J.; YE, X. Cellulase assays. Methods Mol Biol 2009, 581: 213–231.

[20] MANDELS, M.; ANDREOTTI, R.; ROCHE, C. Measurement of saccharifying cellulase. Biotechnol Bioeng Symp 1976, 6:21–33.

[21]   BAILEY, M. J.; NEVALAINEN, K. M. H. Induction, isolation and testing of stable *Trichoderma reesei* mutants with improved production of solubilizing cellulose. Enzyme Microb Technol 1981, 3: 153-162.

[22]   COWARD-KELLY, G.; AIELLO-MAZZARI, C.; KIM S, GRANDA C.; HOLTZAPPLE, M. Suggested improvements to the standard filter paper assay used to measure cellulase activity. Biotechnol Bioeng 2003, 82:745–9.

[23]   ADNEY, B.; BAKER, J. Measurement of Cellulase Activities. LAP-006 NREL Analytical Procedure, National Renewable Energy Laboratory, Golden, CO, 1996.

[24]   EVELEIGH, D. E.; MANDELS, M.; ANDREOTTI, R.; ROCHE, C. Masurement of saccharifying cellulase. Biotechnology for Biofuels 2009, 2:21.

[25]   XIAO, Z.; STORMS, R; TSANG, A. Microplate-based filter paper assay to measure total cellulase activity. Biotechnol Bioeng 2004, 88: 832–837.

[26]   DECKER, S. R.; ADNEY, W. S.; JENNINGS, E., VINZANT, T. B.; HIMMEL, M. E. Automated filter paper assay for determination of cellulase activity. Appl Biochem Biotechnol 2003, 105 – 108: 689–703.

[27]   LYND, L. R.; WEIMER, P. J; VAN ZYL, W. H.; PRETORIUS, I. S. Microbial cellulose utilization: Fundamentals and biotechnology. Microbiol Mol Biol R 2002, 66: 506–77.

[28]   KWON, K. S.; LEE, J.; KANG, H. G.; HAH, Y. C. Detection of 3-glucosidase activity in polyacrylamide gels with esculin as substrate. Appl Environ Microbiol 1994, 60(12): 4584–4586.

CHAPTER 5

Methods for Detection of Amylolytic Activities

Aline Machado de Castro[1,*] and Bernardo Dias Ribeiro[2]

[1]Biotechnology Division, Research and Development Center, PETROBRAS, Rio de Janeiro, Brazil and [2]School of Chemistry, Federal University of Rio de Janeiro, Rio de Janeiro, Brazil

Abstract: Amylases are some of the most important industrial enzymes. Their family is comprised of enzymes with different specificities against a broad range of substrates. The group of endoamylases, which includes α-amylases, catalyzes the cleavage of internal α-1,4 linkages in amylose and amylopectin structures, releasing dextrins of various lengths. Exoamylases, on the other hand, act preferentially in external regions of the substrate, being represented by several enzymes which act towards molecules as small as maltose (*e.g.,* maltase) until polysaccharides (*e.g.,* glucoamylases). The third group comprises debranching amylases, including pullulanases and isoamylases, which act in α-1,6 bonds (branching linkages). Synergy between these groups of enzymes is crucial for the improvement of product release rate. Therefore, it is important to assess methods for the more specific detection as possible for each main group of amylolytic enzymes, in order to understand their synergy and evaluate potential microbial strains for their production. This chapter contains an overview of the mode of action of the main amylolytic enzymes and presents protocols for the quantification of the activity of the three major groups of amylases.

Keywords: Amylase, starch, amylose amylopectin, endoamylases, exoamylases, pullulanase, glycogen, polysaccharides.

INTRODUCTION

Starch is an abundant carbon source on Earth and it is present as energy reserve in almost all higher plant cells [1]. It is constituted by two major classes of homopolysaccharides: amylose, a linear chain of α-1,4-linked glucopyranosidic units; and amylopectin, which comprises the linear α-1,4-linked chains with branching spots (α-1,6 linkages). The latter polysaccharide is the major

Address correspondence to Aline Machado de Castro: Biotechnology Division, Research and Development Center at Petrobras, Brazil; Tel: +55 21 2162-2811; Fax: +55 21 2162-6973; E-mail: alinebio@petrobras.com.br

responsible for the crystallinity of starch granules, due to its cluster structure where highly packed and ordered parallel chains are arranged [2].

There are several amylases which action is related to starch polysaccharides. One group includes glycosyltransferases (EC 2.4.z.x, where z and x depend on the enzyme studied), such as starch synthase (EC 2.4.1.21) and branching enzyme (EC 2.4.1.18), which act on starch biosynthesis pathways [3]. The second group comprises hydrolases (specifically *S-* and *O-* glycosidases, EC 3.2.1.y), which are the amylolytic enzymes presenting the highest industrial interest. Among the amylases classified in this sub-subclass, the three major groups are [4, 5, 6]:

Endoamylases

Amylases that catalyzes the hydrolysis of internal α-1,4 bonds in both amylose and amylopectin chains, with the concomitant release of oligosaccharides of various lenghts. α-amylase (EC 3.2.1.1) is the major endoamylase, and is mainly produced by bacteria (*e.g., Bacillus* sp.) and fungi (*e.g., Aspergillus* sp. and *Rhizopus* sp.) [7]. Enzymes from the former group of microorganisms tend to act best at higher temperatures (80-95°C) [8], whereas endoamylases produced by the fungal strains have their optimal temperature commonly reported as being below 70 °C [9, 10, 11, 12, 13]. The endoamylolytic activity has also been reported as liquefying activity [14] due to the role of α-amylase in bioethanol production from starch [15].

Exoamylases

This group comprises amylases that catalyzes the hydrolysis of terminal glucosidic units, releasing mainly glucose (such as glucoamylase, EC 3.2.1.3; and α-glucosidase, 3.2.1.20) and maltose (β-amylase, EC 3.2.1.2). α-glucosidase (also known as maltase) preferentially acts in soluble and small maltooligosaccharides (including maltose), whereas glucoamylase tends to act best in large chains [4]. Another difference between these two exoamylases is their catalytic mechanism. The former retains the α-anomeric configuration that glucose residues present in the polysaccharide chain. The latter, on the other hand, inverts the configuration of hydroxyl groups linked to carbon 1 in pyranosidic ring, so that the product of hydrolysis is a β-glucose. Glucoamylase are also reported due to their action towards branching linkages (α-1,6), although kinetics is rather slower compared to its catalysis in α-1,4 bonds [16]. So as fungal α-amylases, exoamylases produced by filamentous fungi are also reported to present their optimal catalytic power at temperatures lower than 70 °C [17, 18, 19]. Regarding β-amylase, it is known that it is more commonly found in higher plants, as well as produced by bacteria [6]. In addition, there are some other exoamylases that catalyzes the cleavage of α-1,4 linkages and the release of maltooligosaccharides (*e.g.,* maltotetraose-forming amylase, EC 3.2.1.60, which main product is maltotetraose), but they will not be discussed in details, since they are not the most commonly reported hydrolytic amylases. In a similar manner to endoamylases, the exoamylolytic activity has also been reported as saccharifying activity [14] due to the role of glucoamylase in bioethanol production from starch [15].

Debranching Amylases

This group is formed by amylolytic enzymes capable to act in α-1,6 linkages. The major enzymes are: 1) pullulanase (EC 3.2.1.41), that acts in pullulan and glycogen, as well as towards amylopectin. If the substrate is pullulan, hydrolysis product will be exclusively maltotriose, since α-1,6 linkage in its polysaccharide regularly appears at each three glucose units. This enzyme, additionally to related enzymes (*e.g.,* isopullulanase, EC 3.2.1.57; and amylopullulnase or neopullulanase, EC 3.2.1.135) is mainly produced by mesophilic and thermophylic bacteria [4, 20], and presents optimal activity between 50-100 °C [21]. 2) isoamylase (EC 3.2.1.68), which differs from pullulanase due its inability to act towards pullulan. Hydrolysis in the presence of this latter enzyme releases maltooligosaccharides containing at least two glucose units. 3) Limit dextrinase (EC 3.2.1.142), which action results in the release of oligosaccharides no shorter than maltose, as isoamylase does. Limit dextrinase can act in α– and β–limit dextrins in amylopectin, glycogen and pullulan structures, and is more often found in plants [20].

The mode of action of hydrolytic amylases towards an amylopectin structure is illustrated in Fig. **1**.

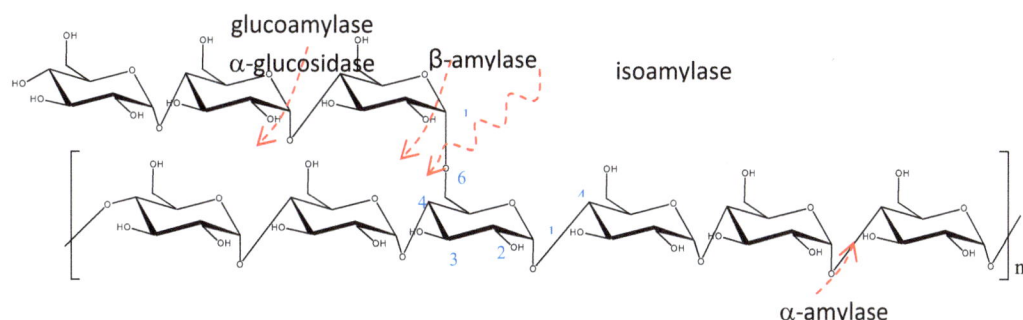

Figure 1: Mode of action of amylases towards an amylopectin structure.

The large variety of amylolytic enzymes has led to the development of several quantification methodologies. Despite the presence of different amylases in a same enzyme extract, the specificity of each one should be explored when certain protocol is applied. To guarantee that only the desired enzyme will act at the method conditions, if the extract was not previously purified, is a hard task. However, based on fundamentals regarding the mode of action of each major amylolytic enzyme, some methodologies can be used for the primarily dosage of one or even few enzyme activities. Table **1** exemplifies the heterogeneity of protocols for the measurement of amylolytic activities.

Table 1: A selection of methods for the measurement of amylolytic activities.

Source of enzyme	Activity measured [1]	Reaction conditions: substrate; t (min); T(°C); pH	Products detection	Activity definition (1 U)	Reference
Fusarium moliniforme	Amylase (total)	5.0 mg starch; 10; 40; 6.0	Starch-iodine complex	Hydrolysis of 0.1 mg starch in 10 min.	[22]
Rhizopus microsporus	Amylase (total)	1% starch; NR; 65; 5.0	Reducing sugars (RS)	μmol RS min^{-1}	[23]
Aspergillus niger	α-amylase	1% soluble starch; 30; 40; 5.0-6.0	Reducing sugars (RS)	μmol RS (maltose equivalent) min^{-1}	[24]
Macrophimina phaseolina	α-amylase	0.5% starch; 20; 50; 5.5	Starch-iodine complex	Hydrolysis of 0.1 mg starch min^{-1}.	[14]
Aspergillus awamori	Exoamylase	1% soluble starch; 10; 40; 5.0	Glucose	μmol glucose min^{-1}	[25]

Table 1: contd...

Penicillium restrictum	Glucoamylase	4% soluble starch; NR; 45; 4.2	Reducing sugars (RS)	μmol RS min^{-1}	[26]
Leucoagaricus gongylophorus	α-glucosidase	2% maltose; 60; 30; 5.0	Glucose	μmol glucose min^{-1}	[27]
Bacillus cereus	pullulanase	1% pullulan; 20; 50; 6.8	Reducing sugars	μmol RS (glucose equivalent) min^{-1}	[28]
Bacillus sp.	pullulanase	1% pullulan; 30; 50; 6.0	Maltotriose (chromatography)	μmol maltotriose min^{-1}	[29]
Musa acuminata (banana)	isoamylase	1% β-dextrin limit; 30; 30; 7.4	Reducing sugars	NR	[30]

[1] According to the correspondent authors
NR: Data not reported

Furthermore, it is important to notice that the right methodology to measure an amylolytic activity depends on the nature of the sample and the target applications for the biocatalyst. It is recommended that the prospective protocol be validated, using a standard sample with similar characteristics compared to the samples which activities need to be determined. Validation includes: comprehension of the range of product released (based on the method for its detection) which corresponds to the action of the enzyme under initial rate of hydrolysis; and selection of the proper reaction conditions, such as pH and temperature (a "kick off" can be based on literature with enzymes produced by similar species, or based on screening trials for the specific strain that will be used to generate the samples with unknown activities).

The objective of this chapter is to present some protocols for the detection of the three main groups of amylolytic enzymes. The most specific as possible methods were selected, and are presented below.

Protocols for the Measurement of Amylolytic Activities

Endoamylolytic Activity

Principle of the Method

Starch polysaccharides tend to form helical structures, so that, *e.g.,* in amylose, a double helix can be generated by association of two single strands [31]. Within these structures, iodine ions (I_3^-) interact, forming what is named as "starch blue complex". As revised by Tomasik and Schilling [32], several hypothesis have been made regarding the structure of the starch-iodine complex, since its first report in 1814, including: characterization of the complex as a physical mixture, a solid or a colloidal solution of iodine; a polysaccharide-adsorbed iodine complex; and a change in the starch structure caused by iodine, so that the modified starch (without ions adsorbed) becomes blue.

Although starch-iodine interaction has been mostly characterized as blue complexes, it should be noticed that this color is due to the interaction of iodine with amylose. The interaction of iodine with amylopectin results in a purple-red color, with a peak of absorptivity in visible spectrum (548 nm) quite different from the former (630 nm) [33]. However, since the blue color is so stronger that the purple one, the resultant solution shows the prevalence of the interaction iodine-amylose.

Moreover, a concern that has been reported regarding starch-iodine complexes is the role of potassium iodide. Some authors defend its importance in the development of the color, whereas others consider it an unnecessary component, informing that it is not adsorbed by starch at all. Finally, there are some works that report that starch-iodine complexes are formed both in the presence or absence of iodide, and that the difference between these cases is only the color intensity [32].

This principle is used for the determination of starch content in plants [34]. In these cases, additionally to iodine interaction with amylose and amylopectin, color formation by interaction of iodine with xylan can also be observed, but only in the presence of at least 30% $CaCl_2$ [35].

Therefore, the starch hydrolyzed due to biocatalytic action can be estimated by the residual starch in solution. Since helical structures in both amylose and amylopectin are characteristics of long chains [31], the major activity that is measured by this principle is of endoamylases, which are capable to rapidly reduce the degree of polymerization of the polysaccharides.

Procedure

Reagents

1. Fosforic acid;

2. Boric acid;

3. Glacial acetic acid;

4. Sodium hydroxide (pearls);

5. Potato starch (soluble, for iodometry);

6. Enzyme extract;

7. Distilled water;

8. Iodine reagent (0.01 N I_2 in HCl 0.02N solution). This reagent can be bought already prepared from suppliers for clinical analysis laboratories or can be prepared in your laboratory. If the first alternative is chosen, watch the expiration date of the reagent. If you prepare your own reagent, put it in an amber glass flask, labeling properly with solution identification, date of preparation and your name.

Materials

1. Plastic microtubes (2 mL, polypropylene);

2. Pipettes (10 μL, 100 μL, 1 mL and 5 mL);

3. Glass flask (100 mL);

4. Glass cuvette;

5. Spectrophotometer (visible range);

6. Thermostatic bath (preferably digital, 0.1 °C precision);

7.   Vortex mixer;

8.   Chronometer.

Preparation of Solutions

Universal Buffer 120 mM [36]

This buffer is indicated due to its large buffering range (confidence range from 2.0-11.0), so that it can be used for almost all samples from the most diverse sources. The buffer is based on two stock solutions:

A (acid solution): Solution containing 40 mM of fosforic, boric and acetic acids, each, in distilled water.

B (alkali solution): Contains 200 mM sodium hydroxide in distilled water.

Solutions must be mixed in different proportion, depending on the pH desired (Fig. **2**). Bacterial amylases commonly act best in solutions with pH between 6.0 and 7.0, whereas for fungal amylases, the optimal pH is usually in the range 4.0-5.5 [37]. For this protocol, it will be used a buffer solution presenting pH 5.0. When necessary, the pH of the final buffer solution can be adjusted to the exact desired value using 0.1N sodium hydroxide or 0.1N chloridric acid.

Substrate Solution

For the preparation of a soluble starch solution, 0.5 g of potato starch was added to 100 mL of universal buffer pH 5.0. The solution should be heated to about 65 °C for the complete gelatinization, under magnetic stirring. After cooling, the solution is ready to be used for the assays. The solution should be stored refrigerated for no longer than 2 weeks in glass flasks, labeled with date of preparation as well as solution and operator identifications. When you will use this solution, after its maintenance on the refrigerator, possibly you may have to hest it for the complete solubilization of starch.

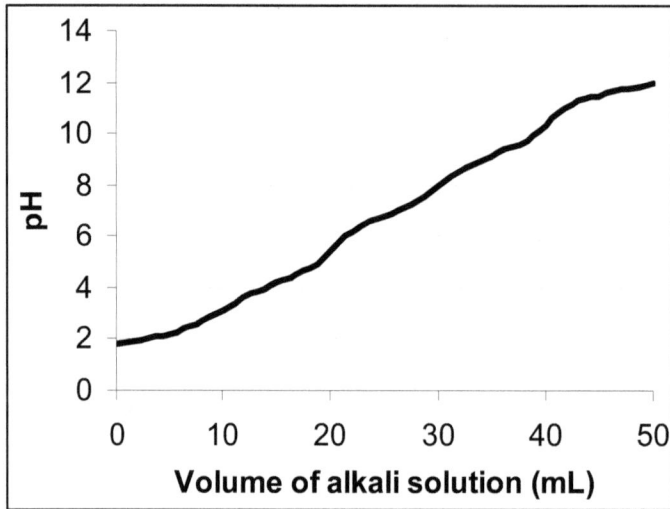

Figure 2: Dependence of the pH of universal buffer solutions to the volume of alkali stock solution. The volume of acid solution should be added in order to complete 50 mL. Source: Elaborated from Britton and Robinson [36].

Operation

Calibration Curve

1. Prepare diluted substrate solutions, by diluting the stock substrate solution with universal buffer 1.5, 2.0, 4.0 and 10.0 times, for a total volume of 500 μL, each solution;

2. Distribute 100 μL of the diluted solutions in plastic microtubes (at least three tubes for each solution) and then add 90 μL of iodine reagent, in order to stop enzymatic reaction. Use in both additions the 100 μL pipette;

3. Cover the flasks and mix vigorously in a vortex mixer for 2 sec.;

4. Carefully open the flasks and add 1 mL of distilled water, with the proper pipette;

5. Cover the flasks again and mix them vigorously for more 2 sec.;

6. A blank, containing 100 μL of distilled water instead of substrate solution, must be prepared at the same time of the standard samples;

7. Transfer 1 mL of final blank solution to the cuvette and add more 4 mL of distilled water in the cuvette, using a 5 mL pipette;

8. Mix the solution by pushing and pulling the liquid;

9. Then, insert the filled cuvette in the proper device inside a spectrophotometer and calibrate the equipment, making this first absorbance as zero. Repeat steps 7-9 for the standard samples (including those which were not diluted) and register the absorbance values of each replicate. Construct a calibration curve plotting the mean absorbance values observed for each substrate solution *vs.* the total starch amount in the sample (in mg). Check the coefficient of determination. It should be higher than 0.9900 for accurate determination of starch content in samples with unknown activities.

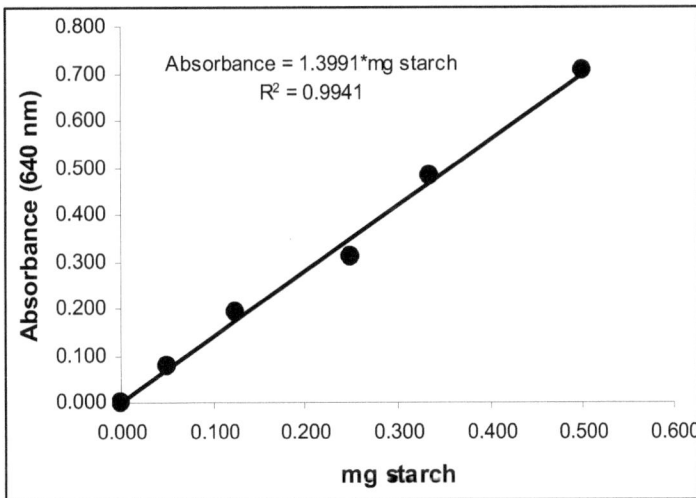

Figure 3: Example of a calibration curve for quantification of endoamylolytic activity.

Assay of Enzyme Extracts

For the determination of endoamylolytic activity in enzyme extracts with unknown activities, distribute 10 μL of the samples in plastic test tubes, using the 10 μL pipette, for more precision. If you expect high activities in the samples, a dilution may be necessary. In this case, dilute the stock sample using universal buffer pH 5.0. Add 90 μL of substrate solution to each tube and immediately incubate them in a thermostatic bath set at 40 °C. After exactly 3 min, remove the tubes from the bath and add 90 μL of iodine reagent in each one. Since the reaction time proposed is short (note that this method was previously validated for fungal endoamylases and that this is a reaction time safe to ensure that the enzymes, when properly diluted, are acting in initial rate of hydrolysis), the time required to add the reagent in several samples can be long enough to make difference in the reaction conditions. If you have many samples, therefore, make the following: once all enzyme extracts are distributed in test tubes (at least in triplicate), stand up in front of the thermostatic bath with a chronometer and start counting time. When the time reaches 15 sec, add the substrate solution in the first tube and immediately incubate it in the bath. Repeat this procedure with the subsequent tubes, adding substrate every 15 sec, until 2 min and 45 sec from the beginning of the time count. At this time, change the pipette tip and, when the time in the chronometer reaches 3 min and 15 sec, add 90 μL of iodine reagent in the first tube incubated in the bath, and do the same in the following tubes every 15 sec. With this procedure, you guarantee that the time of incubation of each tube will be exactly 3 min. Do steps 4 and 5 described in the calibration curve protocol. Each sample will have a blank, where the solutions must be added in the following order: 10 μL of enzyme extract (presenting the same dilution as the sample), 90 μL of iodine reagent and then 90 μL of substrate solution. Adopting this sequence, you will ensure that no biocatalysis will occur towards starch in the blank sample. Repeat steps 7-9 for the reacted as well as the blank samples and register the absorbance values of each replicate.

Calculation of Endoamylolytic Activity

For the calculation of activity, use Equation 1. One unit of endoamylolytic activity is defined as representing the amount of enzyme that catalyzes the hydrolysis of 1.0 mg of starch per minute, under the assay conditions.

$$\text{Activity (U mL}^{-1}) = \frac{\alpha \times \left(\text{abs}_{\text{blank}} - \text{abs}_{\text{sample}} \right) \times f_{\text{d}}}{3 \times 0.01} \qquad (1)$$

Where: α is the angular coefficient of the calibration curve (in the example shown in Fig. **3**, $\alpha=1.3991$); abs is the absorbance values registered for reacted samples and blanks; f_{d} is the dilution factor (in universal buffer) for the samples and blank; 3 represents the time of enzymatic reaction (min); and 0.01 represents the volume of enzyme extract (mL) used for the assay.

Exoamylolytic Activity

Principle of the Method

As described previously, exoamylases act mostly in terminal residues in amylose and amylopectin chains, releasing glucose and maltose as main products [4]. Several studies [17, 26, 38, 39, 40, 41, 42] adopt for the determination of exoamylolytic activity (mainly regarding glucoamylase) the method described by Miller [43], which is based on the detection of the total reducing power of sugars (both glucose and maltooligosaccharides) towards a 3,5-dinitrosalicylic acid-based reagent [44]. However, this procedure is not the most appropriate for the detection of exoamylases, since even endoamylases (*e.g.*, α-amylase) can act for the release of maltooligosaccharides [45].

For a more specific estimation of exoamylolytic activity it is recommended, therefore, to use a method for detection of only glucose, as reported by [46], [18], [19] and [25]. This can include the use of high performance liquid chromatography or the use of spectrophotometry, based on an enzymatic assay. This latter is the most reported methodology, due to its feasible, simple and cheap characteristics. This detection is based in the use of a reagent containing the enzymes glucose oxidase (GOD, EC 1.1.3.4) and horseradish peroxidase (POD, EC 1.11.1.7) [47, 48]. The former enzyme oxidizes glucose, forming hydrogen peroxide, whereas the latter biocatalyst oxidizes a proton donor in the presence of the peroxide, resulting in the release of a redox coupled indicator [20]. There are several indicators that can be used in this second reaction, such as *o*-dianisidine, 4-aminoantipyrine associated with aromatic compounds, and 3-methyl-2-

benzothiazolinone hydrazone associated with 3-methylaminobenzoic acid. Depending on the redox coupled indicator used, the peak of absorptivity can vary, and in the majority of cases, ranges from 460-610 nm [49]. Reactions 1 and 2 represent the oxidation of glucose, and the formation of a redox coupled indicator from 4-aminofenazone (4-AF) and phenol, respectively.

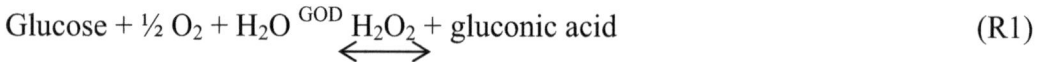

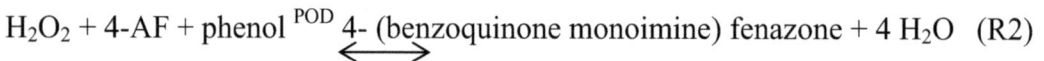

$$\text{Glucose} + \tfrac{1}{2}\,O_2 + H_2O \overset{\text{GOD}}{\longleftrightarrow} H_2O_2 + \text{gluconic acid} \qquad\qquad (R1)$$

$$H_2O_2 + 4\text{-AF} + \text{phenol} \overset{\text{POD}}{\longleftrightarrow} 4\text{- (benzoquinone monoimine) fenazone} + 4\,H_2O \quad (R2)$$

The glucose released by action of exoamylases can be quantified, therefore, by using an enzymatic method, with colorimetric determination of the final product.

Procedure

Reagents

1. Fosforic acid;

2. Boric acid;

3. Glacial acetic acid;

4. Sodium hydroxide (pearls);

5. Potato starch (soluble);

6. Glucose (>99% purity);

7. Enzyme extract;

8. Distilled water;

9. Kit containing glucose oxidase/ horseradish peroxidase (GOD/POD) enzymes;

   This reagent can be bought already prepared from suppliers for clinical analysis laboratories or can be prepared in your laboratory. If the first alternative is chosen, watch the expiration date of the reagent. The kit proposed in this protocol comprises:

- A reagent containing 25 mM of 4-AF in buffered solution (Tris 920 mM);

- A reagent containing 55 mM of phenol; and

- An enzymatic solution containing at least 1 KU mL^{-1} of GOD and 0.15 KU mL^{-1} of POD.

If you prepare your own reagent, put it in an amber glass flask, labeling properly with solution identification, date of preparation and your name.

Materials and Equipment

1. Plastic microtubes (2mL, polypropylene);

2. Pipettes (10 µL, 100 µL and 1 mL);

3. Glass flask (100 mL);

4. Graduate beaker (100 mL);

5. Glass cuvette;

6. Spectrophotometer (visible range);

7. Thermostatic bath (preferably digital, 0.1 °C precision);

8. Water boiling bath;

9. Cold water bath;

10. Plastic sealing caps for microtubes (only if they do not present screw caps);

11. Vortex mixer;

12. Chronometer.

Preparation of Solutions

Universal Buffer 120 mM

This solution must be prepared as described in the protocol for determination of endoamylolytic activity.

Glucose Standard Solution

Dissolve 50 mg of pure D-glucose in 100 mL of distilled water. Distribute the solution in several plastic tubes (*e.g.,* 2 mL) properly labeled and stock them frozen.

Substrate Solution

For the preparation of a soluble starch solution, 1.0 g of potato starch was added to 100 mL of universal buffer pH 5.0. The solution should be heated to about 65 °C for the complete gelatinization, under magnetic stirring. After cooling, the solution is ready to be used for the assays. The solution should be stored refrigerated for no longer than 2 weeks in glass flasks, labeled with date of preparation as well as solution and operator identifications. When you will use this solution, after its maintenance on the refrigerator, possibly you may have to hest it for the complete solubilization of starch.

GOD/POD Reagent

For the preparation of 100 mL of this reagent, add, in the following order: 70 mL of distilled water, 5 mL of reagent solution 1 (4-AF) of the enzymatic kit, 5 mL of reagent solution 2 (phenol) of the enzymatic kit and 300μL of solution 3 (enzymes) of the kit. Then, complete the volume with distilled water. This reagent should be stored refrigerated for a maximum of 4 weeks, in an amber flask.

Operation

Calibration Curve

1. Every time you need to prepare a new calibration curve (if you will perform the assay very often, it is recommended to update the

calibration curve every 4 weeks), withdraw one of the test tubes from the freezer and, after use, discharge the residual volume. Prepare diluted substrate solutions from the glucose standard stock solution, diluting it 1.5, 2.0, 2.5, 5.0 and 10.0 times with distilled water, in order to make a total volume of 500 μL, each solution.

2. Distribute 100 μL of the diluted solutions in plastic test tubes (at least three tubes for each solution) and then add 1 mL of GOD/POD reagent. A blank, containing 100 μL of distilled water instead of substrate solution, must be prepared at the same time of the standard samples.

3. Close the tubes, mix each one by inversion (three times) and incubate them at 37 °C for 15 min.

4. After this reaction, fill the cuvette with the blank sample and put it in the proper device inside a spectrophotometer, calibrating the equipment.

5. Register the absorbance values of each replicate of the standard samples, at 505 nm.

6. Construct a calibration curve plotting the mean absorbance values observed for each substrate solution *vs.* the total amount of μmols of glucose in the sample. Check the coefficient of determination. It should be higher than 0.9900 for accurate determination of starch content in unknown samples. An example of calibration curve for this procedure is shown in Fig. **4**.

Assay of Enzyme Extracts

For the determination of exoamylolytic activity in unknown enzyme extracts, distribute 10 μL of the samples in plastic microtubes, using the 10 μL pipette. As explained for the quantification of endoamylases, if you expect high activities in the samples, a dilution may be necessary. In this case, dilute the stock sample using universal

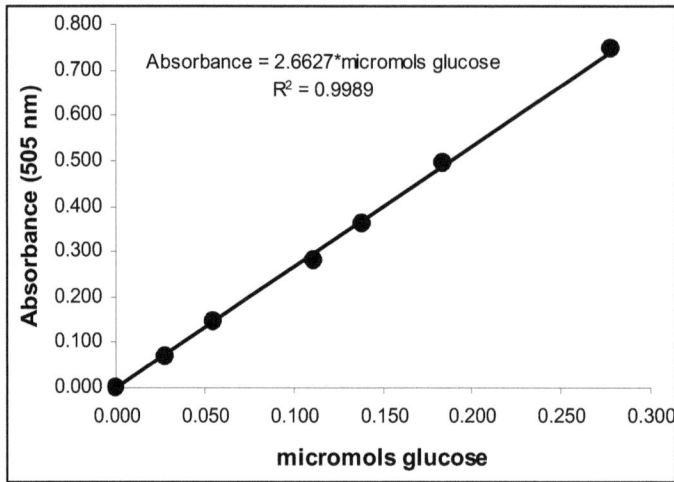

The plot shows: Absorbance = 2.6627*micromols glucose, $R^2 = 0.9989$

Figure 4: Example of a calibration curve for quantification of exoamylolytic activity.

buffer pH 5.0. Add 90 μL of substrate solution to each tube and immediately incubate them in a thermostatic bath set at 40 °C. After exactly 10 min, remove the tubes from the bath, fit sealing caps (if necessary) in each tube and immediately incubate them in a water boiling bath, for 5 min, for the inactivation of the enzymes. Please note that this time is generally enough to inactivate fungal enzymes. Nevertheless, exoamylases from bacterial origin may present high thermal stability [50], so that this time may not be sufficient. In this case, keep the tubes in the water boiling bath for a longer time (10-15 min). You may try to tailor your reaction conditions by doing preliminary trials, with samples incubated in different times, and compare the results to reach a decision. It should be informed also that each sample has to have a blank, which is generated mixing the enzyme extract and the substrate solution and immediately incubating the tube in the water boiling bath. After the inactivation time, put the tubes in a cold water bath and wait some minutes for their cooling. Then, invert the tubes three times, so that the liquid fraction in the bottom of the tubes will be mixed with the condensed water in their caps. Open the tubes and add 1 mL of GOD/POD reagent in each. Close the tubes and incubate them for 15 min at 37°C, as described for the standard samples previously. After this time, open the tubes and start registering their absorbance values (at 505 nm), remembering that the blanks must be used to zero the equipment before each correspondent sample.

Calculation of Exoamylolytic Activity

For the calculation of exoamylolytic activity, use Equation 2. One unit of exoamylolytic activity is defined as representing the amount of enzyme that catalyzes the releasing of 1.0 μmol of glucose per minute, under the assay conditions.

$$\text{Activity (U mL}^{-1}) = \frac{\alpha \times \text{abs} \times f_d}{10 \times 0.01} \qquad (2)$$

Where: α is the angular coefficient of the calibration curve (in the example shown in Fig. **4**, α=2.6627); abs is the absorbance values registered for reacted samples; fd is the dilution factor (in universal buffer) for the samples; 10 represents the time of amylolytic reaction (min); and 0.01 represents the volume of enzyme extract (mL) used for the assay.

Debraching Activity

Principle of the Method

Debranching amylases can be determined by some traditional methods as incubating the enzyme, isoamylase or pullulanase, with the substrate (glycogen, amylopectin or pullulan) under standard conditions (pH = 3.5 - 5, T = 40 – 50°C) for 30 minutes, and measuring the reducing sugars released by Somogyi method [51] or dinitrosalicylic acid (DNS) method [52]. Another method to determine the isoamylase activity is the addition of a solution of iodine ($I_2 - KI$) to amylopectin or soluble waxy starch, which the change in absorbance at 610 nm represents the degree of hydrolysis of the substrate [53, 54]. However these methods are unspecific and must not be used if debranching amylases are associated with others amylases [28, 55].

An alternative detection method that differs from spectrophotometry is high performance liquid chromatography (HPLC) with refractive index detector or evaporative light-scattering detector (ELSD), which can be used for the determination of the degree of hydrolysis by measuring product formation (oligosaccharides and mainly maltotriose). The columns used for sugar analysis are: aminopropylsiloxane-bonded silica; modified silica with ethylenediamine, 1,4-diaminobutane, tetraethylenepentamine (TEPA) and piperazine; silica-based packing carrying bonded amide, cyano, diol, polyol or cyclodextrin; and sulfonated poly(styrene-divinylbenzene) (PS-DVB) copolymers [56].

For separations on aminoalkylsiloxane-bonded silica columns, acetonitrile-water mixtures are usually employed as the mobile phase. The proportion of acetonitrile ranges from 80 to 90% by volume for chromatography of sugars and other carbohydrates of low molecular weight. For chromatographic analysis of oligosaccharides with degree of polymerization (dp) above 3, mobile phases containing acetonitrile in lower proportions are required. Under these conditions, separations are currently carried out at room temperature and with flow rates from 1 to 2 mL min^{-1} [29, 56].

Separation of carbohydrates and alditols in column packed with a cation-exchanger PS-DVB resin can be affected by the degree of cross-linking and type of counterion. Cation exchange resins carrying calcium ions are the most commonly used for chromatography of sugars allowing good separations between monomeric sugars (pentoses, hexoses), alditols, di- and trisaccharides. Carbohydrate separations on cation exchange resins loaded with various metal ions can be carried out with deionized water as the mobile phase, and usually performed with a column temperature ranging from 80 to 85°C, although lower temperatures (65°C) may be used. By heating the HPLC column in the calcium form to 85°C, the rate of interconversion between the α and β anomers of glucose is accelerated, and it is eluted as a single peak [56, 57, 58].

Therefore, the debranching amylase activity can be determined by formation of maltotriose and other reducing sugars which are separated on a aminopropylsiloxane-bonded silica column using HPLC with a refractive index detector.

Procedure

Reagents

    1.   Pullulan (> 99% purity);

2.   Glacial acetic acid;

3.   Sodium Acetate (tri-hydrate);

4.   Enzyme extract containing debranching amylolytic activity;

5.   Acetonitrile;

6.   Bidistilled and Degassed Water;

7.   Maltotriose (> 99% purity)

Materials and Equipment

1.   Test tubes;

2.   Chronometer;

3.   Thermometer;

4.   Refrigerated thermostatic bath;

5.   Pipettes of 1 and 5 mL;

6.   Graduated cylinder of 50 mL;

7.   HPLC with a refractive index detector.

Preparation of Solutions

Acetate Buffer 0.1M, pH 6.0

Make up the following solutions:

A: 0.1M Acetic acid (6.00 g in 1000 mL distilled water).

B: 0.1M Sodium acetate tri-hydrate (13.6 g in 1000 mL distilled water).

Mix 5.22 mL of solution A with 94.78 mL of solution B to get 100 mL of the buffer at the pH 6.0. Check in a pH meter and if the pH deviates from the desired value, slowly add 7M acetic acid or 7M sodium hydroxide solution to reach pH 6.0.

Pullulan Solution 1% in Acetate Buffer 50mM pH 6.0

For the obtainment of 100 mL of solution: weigh 1 g of pullulan and dissolve in 25 mL of NaOH 1N solution under agitation. Add HCl 2 N solution until pH 6.0 and adjust the volume to 50 mL with H_2O. Then, add 50 mL of acetate buffer 0.1M pH 6.0. Store the solution in a glass flask, place the date labeling, the solution name, the name of who performed the operation, and keep in the refrigerator.

Operation

Enzymatic Assay

The reaction mixture consists of 2.0 ml of pullulan (1% w/v) solution in 50 mM sodium acetate buffer (pH 6.0) and 0.5 ml of properly diluted enzyme source. After incubation at 50°C for 30 min, the reaction must be stopped by cooling the tubes in an ice bath and the reducing sugars released by enzymatic hydrolysis of pullulan are then analyzed by HPLC.

Chromatographic Conditions

The mobile phase used is acetonitrile/water (78:22) in isocratic system, previously degassed in an ultrasonic bath. All the aqueous solutions are passed through a 0.2-μm glass filter.

Separation is performed at 35°C (oven temperature) with a flow rate of 0.5 mL min^{-1}, if dimensions of chromatographic column are 100 x 4.6 mm (length *versus* internal diameter). The peak area is converted to maltotriose amount using a calibration curve prepared by injecting 10 μL (the same amount as the sample extract) of the maltotriose solutions (1–10 g L^{-1}).

Calculation of Debranching Amylase Activity (DAA)

One unit of DAA is defined as the amount of enzyme that catalyzes the releasing of 1 μmol of maltotriose equivalent per minute, at 50 °C

(Kunamneni and Singh, 2006). The calculations for DAA activity are shown in Equation 3.

$$DAA\ (UI\ mL^{-1}) = \frac{Dilution*Concentration_{MALTOTRIOSE}\ (\mu mol\ mL^{-1})*Volume_{TOTAL}\ (mL)}{time\ (min)*Volume_{ENZYME\ EXTRACT}\ (mL)} \tag{3}$$

For this calculation, the determination of maltotriose concentration is obtained by correlation with the peak areas in the HPLC chromatograms correspondent to the standard solutions (standard curve).

ACKNOWLEDGMENTS

The authors thank Maria Cristina Saba, Gina Vazquez Sebastian and PETROBRAS for their support for the chapter preparation.

CONFLICT OF INTEREST

The authors confirm that this chapter content has no conflict of interest.

REFERENCES

[1]   SHANNON, J.C.; GARWOOD, D.L.; BOYER, C.D. Genetics and physiology of starch development. In: BeMiller, J.; Whistler, R. (Eds.) Starch – Chemistry and Technology. 3rd Edition. Academic Press. New York. 2009.

[2]   OATES, C.G. Towards an understanding of starch granule structure and hydrolysis. Trends in Food Science & Technology. 1997; v. 8, p. 375-382.

[3]   PREISS, J. Biochemistry and molecular biology of starch biosynthesis. In: BeMiller, J.; Whistler, R. (Eds.) Starch – Chemistry and Technology. 3rd Edition. Academic Press. New York. 2009.

[4]   VAN DER MAAREL, M. J. E. C.; VAN DER VEEN, B.; UITDEHAAG, J. C. M.; LEEMHUIS, H.; DIJKHUIZEN, L. Properties and applications of starch-converting enzymes of the α-amylase family. Journal of Biotechnology. 2002; v. 94, p. 137-155.

[5]   CAZy (Carbohydrate-Active Enzymes Database). Available at: www.cazy.org. Accessed in 09.07.2010.

[6]   CANTAREL, B. L.; COUTINHO, P. M.; RANCUREL, C.; BERNARD, T.; LOMBARD, V.; HENRISSAT, B. The Carbohydrate-Active enzymes database (CAZy): an expert resource for Glycogenomics. Nucleic Acids Research. 2009; v. 37, p. D233-238.

[7]   PANDEY, A.; NIGAM, P.; SOCCOL, C.R.; SOCCOL, V.T.; SINGH, D.; MOHAN, R. Advances in microbial amylases. Biotechnology and Applied Biochemistry. 2000; v. 31, p. 135-152.

[8]   UHLIG, H. Industrial Enzymes and Their Applications. New York: John Wiley & Sons Inc., 1998. 472 p.

[9]   YAGAR, H.; ERTAN, F.; BALKAN, B. Comparision of some properties of free and immobilized α-amylase by *Aspergillus sclerotiorum* in calcium alginate gel beads. Preparative Biochemisty and Biotechnology. 2008; v. 38, n. 1, p. 13-23.

[10]   SUGANUMA, T.; FUJITA, K.; KITAHARA, K. Some distinghishable properties between acid-stable and neutral types of α-amylases from acid-producing *Koji*. Journal of Bioscience and Bioengineering. 2007; v. 104, n. 5, p. 353-362.

[11]   RAMACHANDRAN, S.; PATEL, A.K.; NAMPHOOTHIRI, K.M.; CHANDRAN, S.; SZAKACS, G.; SOCCOL, C.R.; PANDEY, A. Alpha amylase from a fungal culture grown on oil cakes and its properties. Brazilian Archives of Biology and Technology. 2004; v. 47, n. 2, p. 309-317.

[12]   AQUINO, A.C.; JORGE, J.A.; TERENZI, H.F.; POLIZELI, M.L. Studies on a thermostable alpha-amylase from the thermophilic fungus *Scytalidium thermophilum*. Applied Microbiology and Biotechnology. 2003; v. 61, n. 4, p. 323-328.

[13]   ERTAN, F.; Yagar, H.; Balkan, B. Some properties of free and immobilized α-amylase from *Penicillium griseofulvum* by solid state fermentation. Preparative Biochemistry and Biotechnology. 2006; v. 36, n. 1, p. 81-91.

[14]   FERNANDES, L. P.; ULHOA, C. J.; ASQUIERI, E. R.; MONTEIRO, V. N. Produção de amilases pelo fungo *Macrophomina phaseolina*. Revista Eletrônica de Farmácia. 2007; v. IV, n. 1, p. 43-51.

[15]   BOTHAST, R. J.; SCHLICHER, M. A. Biotechnological processes for conversion of corn into ethanol. Applied Microbiology and Biotechnology. 2005; v. 67, p. 19-25.

[16]   NOROUZIAN, D.; AKBARZADEH, A.; SCHARER, J.M.; YOUNG, M.M. Fungal glucoamylases. Biotechnology Advances. 2006; v. 24, p. 80-85.

[17]   ANTO, H.; TRIVEDI, U.B.; PATEL, K.C. Glucoamylase production by solid-state fermentation using rice flake manufacturing waste products as substrate. Bioresource Technology. 2006; v. 97, p. 1161-1166.

[18]   BHATTI H, N.; RASHID, M. H.; NAWAZ, R.; ASGHER, M.; PERVEEN, R.; JABBAR, A. Purification and characterization of novel glucoamylase from *Fusarium solani*. Food Chemistry. 2007; v. 103, p. 338-343.

[19]   RIAZ, M.; PERVEEN, R.; JAVED, M.R.; NADEEM, H.; RASHID, M.H. Kinetic and thermodynamic properties of novel glucoamylase from *Humicola* sp. Enzyme and Microbial Technology. 2007; v. 41, p. 558-564.

[20]   CHANG, A.; SCHEER, M.; GROTE, A.; SCHOMBURG, I.; SCHOMBURG, D. BRENDA, AMENDA AND FRENDA. The enzyme information system: new content and tools in 2009. Nucleic Acids Research. 2009; v. 37, p. D588-592.

[21]   DOMAN-PYTKA, M.; BARDOWSKI, J. Pullulan degrading enzymes of bacterial origin. Critical Reviews in Microbiology, 2004; v. 30, p. 107-121.

[22]   FIGUEIRA, E. L. Z.; HIROOKA, E. Y. Culture medium for amylase production by toxigenic fungi. Brazilian Archives of Biology and Technology. 2000; v. 43, n. 5, p. 461-467.

[23]   PEIXOTO, S.C.; JORGE, J.A.; TERENZI, H.F.; POLIZELI, M.L.T.M. *Rhizopus microsporus* var. *rhizopodiformis*: a thermotolerant fungus with potential for production of thermostable amylases. International Microbiology. 2003; v. 6, p. 269-273.

[24]   DJEKRIF-DAKHMOUCHE, S.; GHERIBI-AOULMI, Z.; MERAIHI, Z.; BENNAMOUN, L. Application of a statistical design to the optimization of culture medium for α-amylase production by *Aspergillus niger* ATCC 16404 grown on orange waste powder. Journal of Food Engineering. 2006; v. 73, p. 190-197.

[25]   CASTRO, A. M.; ANDREA, T. V.; CASTILHO, L. R.; FREIRE, D. M. G. Use of mesophilic fungal amylases produced by solid-state fermentation in the cold hydrolysis of raw babassu cake starch. Applied Biochemistry and Biotechnology. 2010; v. 162, p. 1612-1625.

[26]   GOMBERT, A. K.; PINTO, A. L.; CASTILHO, L. R.; FREIRE, D. M. G. Lipase production by *Penicillium restrictum* in solid-state fermentation using babassu oil cake as substrate. Process Biochemistry. 1999; v. 35, p. 85-90.

[27]   SILVA, A.; BACCI JR., M.; PAGNOCCA, F.C.; BUENO, O.C.; HEBLING, M.J.A. Starch metabolism in *Leucoagaricus gongylophorus*, the symbiotic fungus of leaf-cutting ants. Microbiological Research. 2006; v. 161, p. 299-303.

[28]   NAIR, S.U.; SINGHAL, R.S.; KAMAT, M.Y. Induction of pullulanase production in *Bacillus cereus* FDTA-13. Bioresource Technology. 2007; v. 98, p. 856-859.

[29]   KUNAMNENI, A.; SINGH, S. Improved high thermal stability of pullulanase from a newly isolated thermophilic *Bacillus* sp. AN-7. Enzyme and Microbial Technology. 2006; v. 39, p. 1399-1404.

[30]   BIERHALS, J. D.; LAJOLO, F. M.; CORDENUNSI, B. R.; NASCIMENTO, J. R. O. Activity, cloning, and expression of na isoamylase-type starch-debranching enzyme from banana fruit. Journal of Agricultural and Food Chemistry. 2004; v. 52, p. 7412-7418.

[31]   PÉREZ, S.; BALDWIN, P.M.; GALLANT, D.J. IN: BEMILLER, J.; WHISTLER, R. (Eds.) Starch – Chemistry and Technology. 3rd Edition. Academic Press. New York. 2009.

[32]   TOMASIK, P.; SCHILLING, C. Complexes of starch with inorganic guests. In: Horton, D. Advances in Carbohydrate Chemistry and Biochemistry. V. 53. Academic Press. New York. 1998.

[33]   JARVIS, C.E.; WALKER, J.R.L. Simultaneous, rapid, spectrophotometric determination of total starch, amylose and amylopectin. Journal of Science of Food and Agriculture. 1993; v. 63, p. 53-57.

[34]   AOAC (Association of Official Analyst Chemists). Official Method 948.02 – Starch in plants (titrimetric method). p. 3.5.11. 1962

[35]   YU, X.; ATALLA, R.H. The complex of xylan and iodine: the induction and detection of nanoscale order. Carbohydrate Research. 2005; v. 340, p. 981-988.

[36]   BRITTON, H. T. S.; ROBINSON, R. A. The use of the antimony-antimonous oxide electrode in the determination of the concentration of hydrogen ions and in potentiometric titrations. The Prideaux-Ward universal buffer mixture. Journal of the Chemical Society. 1931; p. 458-478.

[37]   GODFREY, T.; WEST, S. Industrial Enzymology. London: The Macmillan Press Ltd. 1996. 624 p.

[38]   MAMO, G.; GESSESSE, A. Production of raw-starch digesting amyloglucosidase by *Aspergillus* sp. GP-21 in solid state fermentation. Journal of Industrial Microbiology and Biotechnology. 1999; v. 22, p. 622-626.

[39]   MICHELIN, M.; RULLER, R.; WARD, R.J.; MORAES, L.A.B.; JORGE, J.A.; TERENZI, H.F.; POLIZELI, M.L.T.M. Purification and biochemical characterization of a thermostable extracellular glucoamylase produced by the thermotolerant fungus *Paecilomyces variotii*. Journal of Industrial Microbiology and Biotechnology. 2008; V. 35, p. 17-25.

[40]   MARLIDA, Y.; SAARI, N.; HASSAN, Z.; RADU, S.; BAKAR, J. Purification and characterization of sago starch-degrading glucoamylase from *Acremonium* sp. endophytic fungus. Food Chemistry. 2000; v. 71, p. 221-227.

[41]   KAREEM, S.O.; AKPAN, I.; ODUNTAN, S.B. Cowpea waste: a novel substrate for solid state production of amylase by *Aspergillus oryzae*. African Journal of Microbiology Research. 2009; v. 3, n. 12, p. 974-977.

[42]   COSTA, J. A. V.; COLLA, E.; MAGAGNIN, G.; SANTOS, L. O.; VENDRUSCOLO, M.; BERTOLIN, T. E. Simultaneous amyloglucosidase and exo-polygalacturonase production by *Aspergillus niger* using solid-state fermentation. Brazilian Archives of Biology and Technology. 2007; v. 50, n. 5, p. 759-766.
[43]   MILLER, G.L. Use of dinitrosalicylic acid reagent for determination of reducing sugar. Analytical Chemistry. 1959; v. 31, n. 3, p. 426-428.
[44]   SUMNER, J.B. Dinitrosalicylic acid: A reagent for the estimation of sugar in normal and diabetic urine. The Journal of Biological Chemistry. 1921; v. 47, p. 5-9.
[45]   MATSUBARA, T.; AMMAR, Y.B.; ANINDYAWATI, T.; YAMAMOTO, S.; ITO, K.; IIZUKA, M. MINAMIURA, N. Degradation of raw starch granules by α-amylase purified from culture of *Aspergillus awamori* KT-11. Journal of Biochemistry and Molecular Biology. 2004; v. 37, p. 422-428.
[46]   NAGASAKA, Y.; KUROSAWA, K.; YOKOTA, A.; TOMITA, F. Purification and properties of the raw-starch-digesting glucoamylases from *Corticium rolfsii*. Applied Microbiology and Biotechnology. 1998; v. 50, p. 323-330.
[47]   FLEMING, I. D.; PEGLER, H. F. The determination of glucose in the presence of maltose and isomaltose by a stable, specific enzymic reagent. The Analyst. v. 88, p. 967-968. 1963.
[48]   LLOYD, J.B.; WHELAN, W.J. An improved method for enzymic determination of glucose in the presence of maltose. Analytical Biochemistry. 1969; v. 30, p. 467-470.
[49]   BLAKE, D. A.; MCLEAN, N. V. A colorimetric assay for the measurement of D-glucose consumption by cultured cells. Analytical Biochemistry. 1989; v. 177, p. 156-160.
[50]   GILL, R. K.; KAUR, J. A thermostable glucoamylase from a thermophilic *Bacillus* sp.: characterization and thermostability. Journal of Industrial Microbiology and Biotechnology. 2004; v. 31, p. 540-543.
[51]   SOMOGYI, M.; Notes on Sugar Determination. J. Biol. Chem. 1952; 195, 19 - 23,
[52]   FAO (Food and Agriculture Organization of The United Nations); Combined Compendium of Food Additive Specifications, Vol.4: Analytical methods, test procedures and laboratory solutions used by and referenced in the food additive specifications. Joint FAO/WHO Expert Committee on Food Additives, Rome, 2006.
[53]   FANG, T. Y.; TSENG, W. C.; YU, C. J.; SHIH, T. Y.; Characterization of the thermophilic isoamylase from the thermophilic archaeon *Sulfolobus solfataricus* ATCC 35092. Journal of Molecular Catalysis B: Enzymatic, 2005; 33, 99–107.
[54]   FAO; Compendium of Food Additive Specifications, Monographs 4. Joint FAO/WHO Expert Committee on Food Additives, Rome, 2007.
[55]   BIJTTEBIER, A.; GOESART, H.; DELCOUR, J. A. Hydrolysis of amylopectin by amylolytic enzymes: structural analysis of the residual amylopectin population. Carbohydrate Research, 2010; v. 345, p. 235-242.
[56]   CORRADINI, C.; LIQUID CHROMATOGRAPHY, CARBOHYDRATES. IN: COOKE, M.; POOLE, C.F.; WILSON, I.D.; ADLARD, E.R.; Encyclopedia of Separation Science, Volume 5, Level III. Academic Press, San Diego, USA, 2000.
[57]   ROY, A.; MESSAOUD, E.B.; BEJAR, S.; Isolation and purification of an acidic pullulanase type II from newly isolated *Bacillus* sp. US149. Enzyme and Microbial Technology, 2003; 33, 720-724.
[58]   TAKATA, H.; KAJIURA, H.; FURUYASHIKI, T.; KAKUTANI, R.; KURIKI, T. Fine structural properties of natural and synthetic glycogens. Carbohydrate Research, 2009; 344, 654-659.

Send Orders for Reprints to reprints@benthamscience.net

CHAPTER 6

Methods to Determine Xylanolytic Activity

Rodrigo P. Nascimento[1,*], Mônica P. Gravina-Oliveira[2] and Rosalie R. R. Coelho[2]

[1]School of Chemistry, Federal University of Rio de Janeiro, Rio de Janeiro, Brazil and [2]Department of General Microbiology, Institute of Microbiology Paulo de Góes, Federal University of Rio de Janeiro, Rio de Janeiro, Brazil

Abstract: Xylanases are enzymes which catalyze the hydrolysis of 1,4-β-D-xylosidic linkages in xylan. In the present chapter several techniques concerning xylanase activity are discussed, specially those involving qualitative and quantitative assays, which are powerful tools used in screening, selection and enzyme production by xylanolytic microorganisms. This chapter also presents methodologies for enzyme characterization including optimum of pH and temperature, and detection of xylanases through the use of polyacrylamide gel electrophoresis.

Keywords: Xylan, xylooligosaccharides, xylanases, xylosidases, qualitative assays, quantitative assays, Congo red, RBB-xylan, enzymatic activity, DNS, dinitrosalicylic acid, ρ-nitrophenyl-β-D-xylopyranoside, effect of temperature and pH, effect of metal ion, zymogram, D-xylose, β-1,4-xylosidic linkage, β-1,4-endoxylanase, endoxylanase.

INTRODUCTION

Xylanases are enzymes classified as hydrolases, subclass of glycosidases (O-glycoside hydrolases, EC 3.2.1.x) which catalyze the hydrolysis of 1,4-β-D-xylosidic linkages in xylan, the main polymer of the hemicellulose group. They are a widespread group of enzymes, involved in the production of D-xylose, a primary carbon source for cell metabolism and in plant cell infection by plant pathogens, being produced by several organisms [1,2]. The official name of the

---

*Address correspondence to Rodrigo P. Nascimento: School of Chemistry at Federal University of Rio de Janeiro, Brazil; Tel: 55 (21) 2562-8863; Fax: (21) 2562-7567; Email: rodrigopires@eq.ufrj.br

Alane Beatriz Vermelho and Sonia Couri (Eds)

main enzyme from xylanase group is endo-1,4-β-xylanase, but commonly used synonymous terms include xylanase, endoxylanase, 1,4-β-D-xylan-xylanohydrolase, endo-1,4-β-D-xylanase, β-1,4-endoxylanase and β-1,4-xylanase, Fig. **1**.

Xylanases are widespread in nature and they have been reported to be present in marine and terrestrial bacteria, rumen and ruminant bacteria, fungi, marine algae, protozoa, snails, crustaceans, insects, and seeds of terrestrial plants [3]. Interestingly, until 1997, xylanolytic enzymes with hyperthermophilic activity had not been detected. However, in 1999, the first report [4] of xylanase production by the thermophilic archaea *Thermococcus zilligii* has shown us the great possibility to explore extreme environments for biotechnological purposes. Other thermophilics were isolated and identified as promising xylanolytic organisms, such as *Thermotoga* sp., *Caldicellulosiruptor* sp., *Thermoascus aurantiacus* and *Clostridium thermocellum* [5-9].

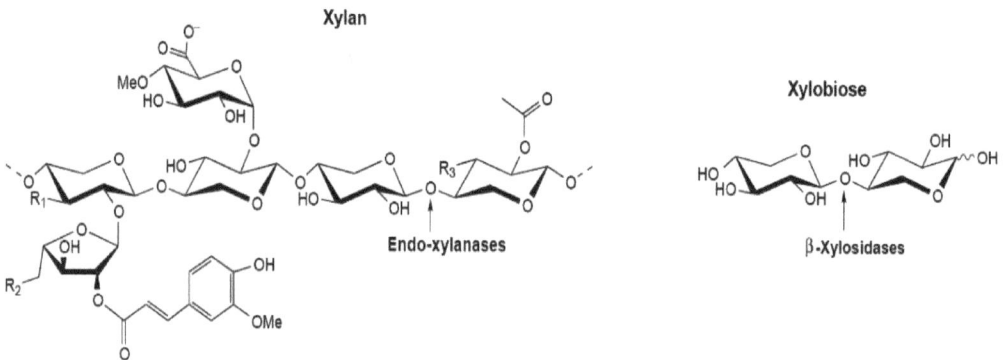

Figure 1: The basic structural components found in hemicellulose and the hemicellulases responsible for degradation of xylan backbone. Endo-xylanase acts randomly on xylan backbone, while xylosidases act on xylobiose and small xylooligosaccharides [10].

Xylanases have potential applications in a wide range of industrial processes, covering all three sectors of industrial enzymes market: food (fruit and vegetable processing, brewing, wine production), feed (animal feeds and baking) and technical industry (pulp and paper, textile, bioremediation/ bioconversion) [2].

Qualitative assays are powerful tools used in preliminary studies aiming at xylan degrading enzyme production. They are particularly useful for screening large numbers of fungal and bacterial isolates for xylanase activity, where a definitive quantitative data is not required. In addition the methodology is straightforward and can be carried out by researchers not specialized in enzymology.

Many strategies to isolate xylan-degrading microorganisms can be adopted, based on the use of selective media containing different types of xylan. The use of oat spelt, birchwood, larchwood, or any other type of xylan as sole carbon source in salt mineral medium can be one of them. In this case xylanase activity can be detected by the visualization of a hydrolysis zone around the microbial growth. Other possibility is the use of a xylan dye complex, Remazol Brilliant Blue-xylan (RBB-xylan), which is a modified (soluble) xylan (4-O-methyl-D-glucourono-D-xylan) bound to the dye Remazol Brilliant Blue R (RBBR) to form the substrate.

Due to its biotechnological importance, the knowledge of the different qualitative and quantitative techniques involving xylanase activity is very important to understand and analyze its biochemical action. Comparison of the enzyme assays used in different applications is hindered by the lack of a standardized method for measurement of xylanase activity. Bailey *et al.* [11] proposed a standardization of methodologies used to determine quantitatively the xylanase activity using different substrates and concentrations. Although the methods used for the assay of *e.g.,* endoxylanase activity are in most cases reported, realistic comparison between the methods is in practice extremely difficult. In the present chapter, the different techniques concerning xylanases activity are discussed, specially for qualitative and quantitative assays which are powerful tools used in screening microorganisms and detecting xylanase activity using methodologies as dinitrosalicylic acid (DNS) method for measuring reducing sugars.

Screening for Xylanolytic Microorganisms

Screening Using Xylan as Substrate

Screening of xylanolytic microorganisms can be accomplished through the use of a selective agar medium containing xylan as sole carbon source. Occasionally other nutrients such as yeast extract can be introduced in the medium, but xylan is always the main substrate. Four different types of xylan can be used, and are commercially available: birchwood, beechwood, larchwood and oat spelts, the four of them being more or less insoluble. The prepared medium is opaque, and after inoculation and incubation, when xylan is depolymerized, hydrolysis zones are formed around microbial growth. Although the four substrates can be used indistinctly, in practice oat spelts xylan gives a better contrast due to its lower solubility [11] which culminates on the formation of xylan aggregates in agar medium. Indeed, according to literature, oat spelts [12-14] and birchwood xylan [15-17] have been used with successful results in screening assays.

In some instances, Congo Red can be used to enhance visualization of the hydrolysis zones. This is a synthesized dye that shows a strong interaction with polysaccharides containing β-(1→4) links and with a significant number of β-(1→3) links, which provides a basis for the assay of β-D-endoxylanase activity [18].

According to literature, the measurement of hydrolysis zones can indicate a semi-quantitative determination of xylanolytic activity through the use of commercial xylanases as standards. However, the diameters of the transparent zones around colonies having low xylanolytic activity can not be measured accurately [19].

Detection of Xylanase Activity Using Oat Spelts Xylan

Objective

To select microorganisms capable of producing endoxylanases based on xylan degradation on agar plate.

Principle

This method is based on the xylan-degrading capacity of microorganisms which produce xylanases. The degradation of the xylan generates a hydrolysis zone around microbial growth on agar plate.

Material

1. Strains of interest;

2. Agar medium containing oat spelts xylan (or other xylan) 1.0 % (w/v) and a mineral salts solution, *e.g.,* [13].

Obs.: Xylan must be used as sole carbon source. Addition of other carbon sources must be done in minimal concentrations.

Procedure

1. Prepare agar medium plates containing oat spelts xylan 1 % (w/v), or other xylan type;

2. Inoculate microorganisms as spots on the agar surface and incubate in adequate conditions. *e.g.*: *Streptomyces malaysiensis* AMT-3 - 30 ºC/ 10 days [13]; Fungi - 30 ºC/ 7 days [20];

3. Detect the xylanase producing microorganisms by a hydrolysis zone surrounding microbial growth as in Fig. **2**.

    Obs.: This test should preferably be carried out using triplicates.

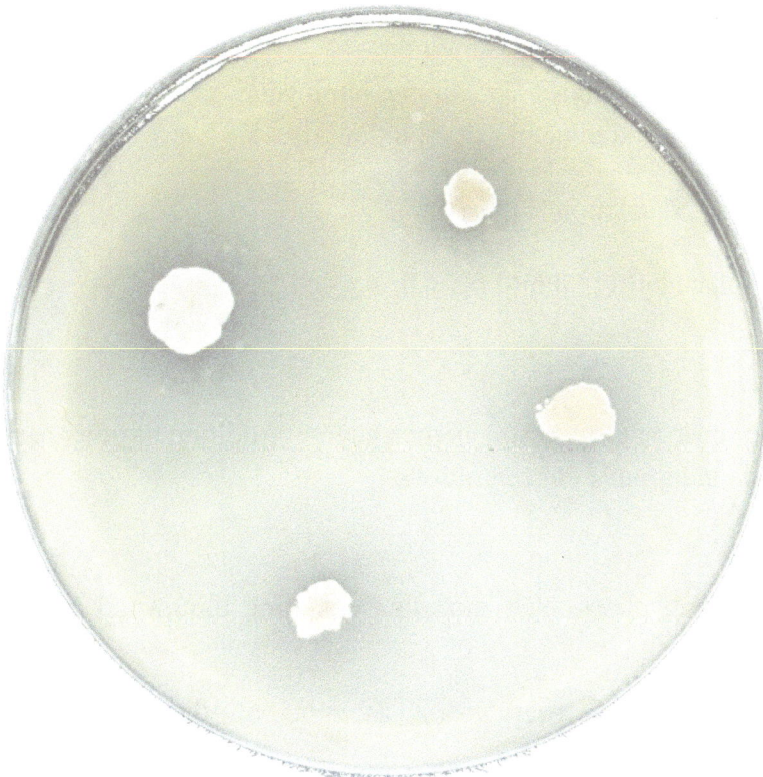

Figure 2: Agar plate with growth of microbial strains in agar medium supplemented with xylan.

Detection of Xylanase Activity Using Birchwood Xylan and Congo Red for Revelation of the Hydrolysis Zones

Objective

To select microorganisms capable of producing endoxylanases based on xylan degradation on agar plate and posterior staining of agar with Congo Red solution.

Principle

This method is based on the xylan-degrading capacity of microorganisms which produce xylanases. The degradation of the xylan generates a hydrolysis zone around microbial growth on agar plate. The interaction between Congo Red and β-(1→4) and β-(1→3) linkages in polysaccharides turns the medium red and thus promotes a better visualization of the hydrolysis zones.

Material

1.   Strains of interest;

2.   Agar medium plates containing birchwood xylan (or other xylan) 0.5% (w/v) and mineral salts solution, *e.g.,* [13];

3.   Congo Red solution 0.1 % (w/v);

4.   Phosphate buffer 50 mM pH 6.0;

5.   NaCl 1 M.

Obs.: Xylan must be used as sole carbon source. Addition of other carbon sources must be done in minimal concentrations.

Procedure

1.   Prepare, in plates, agar medium containing birchwood xylan 0.5 % (w/v);

2.   Inoculate the microorganisms as spots on the agar surface and incubate;

3.   Add Congo Red solution 0.1 % (w/v) during 15 minutes;

4.   Wash the plate with phosphate buffer 50 mM pH 6.0 supplemented with 1 M NaCl to remove the excess of Congo Red;

5.   Detect the xylanase producing microorganisms by the presence of yellow zones of clearance surrounding microbial growth, against a red background area (Fig. **3**) [15, 21].

Obs.: This test should preferably be carried out using triplicates.

Congo Red is also a pH indicator. In acid conditions, Congo Red´s color changes into blue, turning difficult the visualization of hydrolysis zones [22, 23]. For this reason, the use of a buffer with pH 6.0, as described above, might be interesting.

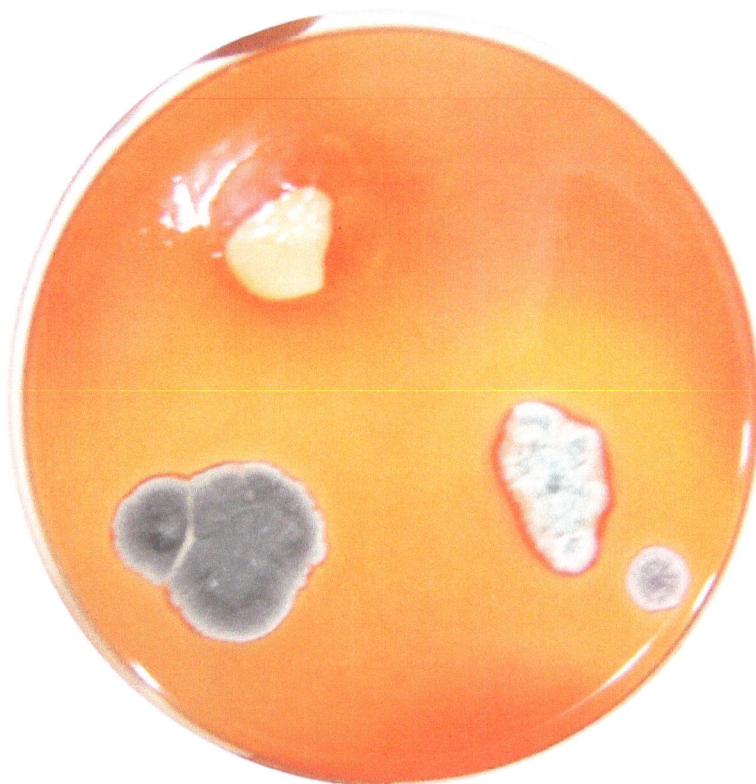

Figure 3: Agar plate with growth of microbial strains in a medium containing xylan, after stained with Congo Red solution, showing the hydrolysis zones generated by xylanolytic activity.

Comments

Despite providing a worse visualization of the hydrolysis zones, when compared to techniques which use dyes (Congo Red) and chromogenic substrates (see below), the use of the first technique, which uses oat spelts xylan, is simpler and inexpensive. On the other hand, the interaction between Congo Red and polysaccharides, like xylan, provides the basis for a sensitive assay system for selection of microorganisms possessing xylanolytic activity. In this case, the technique presents variations concerning to concentration and the type of xylan used in the preparation of the agar medium. The potential advantage of this system for enumeration and characterization of hemicellulolytic microorganisms derives, largely, from the intense color of the dye-xylan complex, which allows the use of low substrate concentrations [18].

Screening with Remazol Brilliant Blue R reagent

Screening methods using a plate containing chromogenic substrate provide a relatively simple way to detect microorganisms with capacity of degrade polysaccharides [24]. The reagent RBBR is a chromogenic substrate, commonly used to verify the presence of enzymatic activity in microorganisms. Some of these enzymes are: ligninases [25], amylases [26], cellulases and also xylanases [27].

To perform a selection of xylan-degrading microorganisms, a substrate containing RBB bound to xylan in agar medium can be used. This substrate can be found as 4-O-methyl-D-glucurono-D-xylan dyed with Remazol Brilliant Blue R, Remazol Brilliant Blue-xylan or simply RBB-xylan (Sigma-Aldrich). In this case the agar medium can be a rich medium supplemented with the RBB-xylan.

Objective

To select xylanolytic microorganisms using solid medium containing xylan covalently bound to the reagent RBB.

Principle

The inoculum of a xylanolytic microorganism in a medium containing RBB linked to xylan causes the generation of a colorless zone around the microbial growth due to the cleavage of xylan and diffusion of dye-labeled oligosaccharides of lower molecular weight in the agar medium [24, 28].

Material

1. Strains of interest;

2. Agar medium containing RBB-xylan 0.2 % (w/v).

    E.g.: Nutrient agar [24], LB agar, Lennox agar [29], Luria Broth Agar [30-31]

Procedure

1. Prepare in Petri dishes the basal medium of choice containing RBB-xylan 0.2 % (w/v);

2. Inoculate the strains of interest as spots on the surface of the agar medium and incubate. *e.g.*: *E. coli* – 37 °C/ over might [5]; Yeasts – 30 °C/ 5 days [32];

3. Detect the xylanase producing microorganisms by the presence of colorless zones surrounding microbial growth, against a blue background area, Fig. **4** [27, 32].

    Obs.: This test should preferably be carried out using triplicates.

Comments

This technique presents variations related to concentrations of the labeled substrate and the medium used. The last one varies according to the assay´s aim (*e.g.*: to screen fungus or bacteria; to select transformed bacteria, *etc.*).

The use of a dye facilitates the visualization of xylan degradation. For this reason, chromogenic assays are preferred techniques in comparison to the ones that use only the polysaccharide to test the presence of the enzymatic activity.

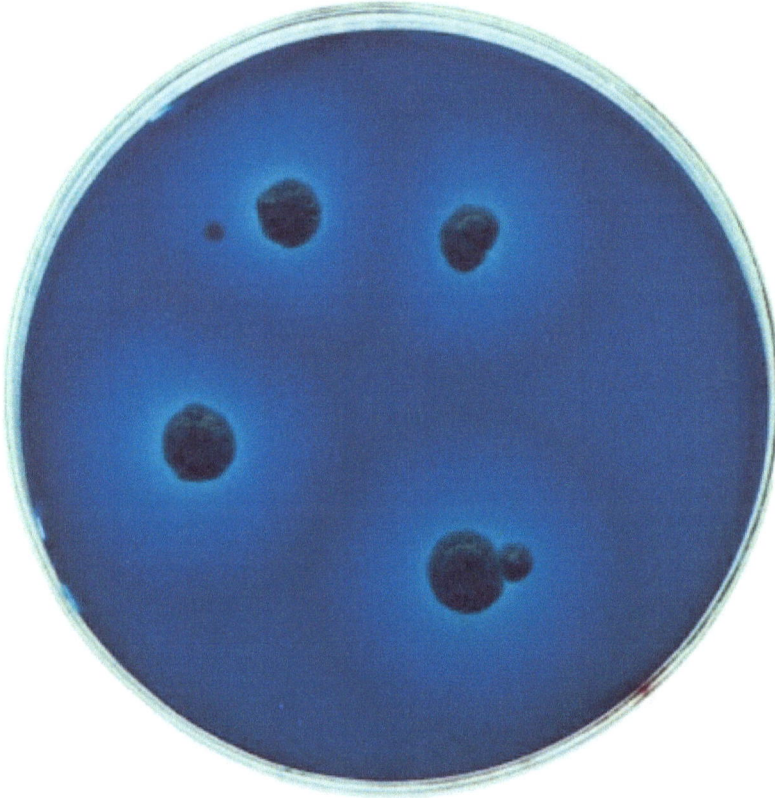

Figure 4: Agar plate with growth of microbial strains in a medium containing RBB-xylan, showing the hydrolysis zones generated by xylanolytic activity.

Enzymatic activity assays

Measurement of Enzymatic Activity Using Xylan as Substrate

Among the colorimetric methods described in literature to measure xylanolytic activity, the methodology proposed by Bailey *et al.* [11] in the article "Interlaboratory testing of methods for assay of xylanase activity" [11] is the most common used. It is based in the use of xylan, and its conversion into xylooligosaccharides. Although the authors have concluded that birchwood xylan was the best substrate, oat spelts xylan [6, 13, 33], as well as beechwood xylan [34, 35] or larchwood xylan has also been used [11].

Objective

Measurement of the enzymatic activity of β-1,4-endoxylanases using xylan.

Principle

The action of β-1,4-endoxylanase on the substrate, xylan, generates reducing sugars (oligosaccharides, in majority) which are measured by the technique of DNS [36]. In a simplest way, the reaction between reducing sugars and DNS is based in the oxidation of the sugars (the C=O groups are oxidized to COOH groups) and reduction of DNS (3,5-dinitrosalycylic acid to 3-amino-5-nitro salicylic acid) with a change in DNS coloration (from yellow to red-orange, or darker, in some cases) which is proportional to the amount of reducing sugars generated.

Material and Equipment

1. Enzyme sample, *e.g.,* as crude extract or supernatant;

2. Birchwood xylan 1 % (w/v) (or other xylan);

3. Sodium citrate buffer 50 mM (pH 5.3);

4. DNS reagent;

5. Sodium hydroxide solution 2 M;

6. Sodium potassium tartrate;

7. Hot plate stirrer;

8. Water bath with temperature controller;

9. Ice bath.

Procedure

Preparing the Substrate Xylan 1.0 % (w/v) [11]

1. Homogenize 1.0 g of xylan in approximately 80 mL of sodium citrate buffer 50 mM (pH 5.3) at 60 °C for 20 minutes and heat to boiling point;

2. Cool with continued stirring, cover and stir slowly overnight;

3. Make up to 100 mL with buffer;

4. Store at 4 °C for a maximum of 1 week or freeze aliquots at -20 °C;

5. Mix with a stirrer after thawing.

Preparing DNS [36]

1. Dissolve 10 g of DNS at room temperature using 500 mL of distilled water;

2. Add 200 mL of sodium hydroxide solution 2 M;

3. After solubilization, add 300 g of sodium potassium tartrate;

4. Bring the volume to 1000 mL with distilled water in a volumetric flask;

5. Store the solution in amber bottle (or dark bottle) protected from light. The standardization must be made between 5 and 10 days after the solution preparation.

Assay (based on Bailey *et al.* [11])

1. Add 750 μL of the substrate solution in a test tube at 50 °C in a water bath;

2. Add 250 μL of the enzyme sample (diluted in citrate buffer 50 mM pH 5.3, if necessary), and mix;

3. Incubate for 20 min at 50 °C.;

4. Immerse the tubes in ice bath;

5. Add 1.5 mL of dinitrosalicylic acid, mix. Boil for 5 min;

6. Immerse the tubes in ice bath;

7. Measure the produced color at 540 nm against the reagent blank [1];

8. Correct the absorbance obtained with the enzyme blank [2] if necessary;

9. Using the standard curve [3], convert the corrected absorbance to enzyme activity units (IU);

10. Calculate activity in the original (undiluted) sample by the following equation:

$$IU/mL^{1-} = \frac{(ABS_{RM} - ABS_{EB}).\,\alpha.\,D}{\Delta T} \qquad \textbf{(1)}$$

α= angular coefficient of the standard linear regression, in $\mu mol.\,mL^{1-}$

ABS_{RM} = mean of absorbance of the reaction mixture (after incubation)

ABS_{EB} = mean of absorbance of enzyme blank (without incubation)

D= enzymatic dilution

ΔT= time of incubation (enzyme-substrate reaction) in minutes

IU= Xylanase international unit is defined as the amount of enzyme required to produce 1 μmol of reducing sugars (xylose) per minute of reaction

Obs.: This test should preferably be carried out using triplicates.

(1) Reagent blank:
✓ 750 μL of xylan 1 % (w/v);
✓ 250 μL of citrate buffer 50 mM pH 5.3;
✓ Incubate 20 min, 50 °C;
✓ 1.5 mL of DNS;
✓ Boil for 5 min and cool;

✓   Measure the color produced at 540 nm.

Obs.: This reaction corresponds to any reducing sugar or color that may be present in the substrate or in any other reagent of the system and that may be subtracted from the result obtained with the enzyme-substrate reaction.

(2)   Enzyme blank:

✓   750 µL of xylan 1 % (w/v);

✓   1.5 mL of DNS;

✓   250 µL of enzyme;

✓   Boil for 5 min and cool;

✓   Measure the color produced at 540 nm.

Obs.: This reaction corresponds to any reducing sugar that may be present in the enzyme sample before the enzyme-substrate reaction occurs, and may be also subtracted. Enzyme blanks are specially required when the sample contains a high level of reducing sugars.

(3)   Xylose standard curve

The standard used in the assay is pure xylose (e.g., Sigma-Aldrich), prepared as a stock solution 0.01 M in sodium citrate buffer 50 mM pH 5.3 (0.15 g per 100 mL of buffer), which can be stored at -20 °C. After thawing, the solutions must be mixed, because they become "layered" on freezing. The stock solution can be diluted according to Tab. 1, however, if necessary, more dilutions can be added.

Table 1: Dilutions and respective xylose concentration used in standard curve

Dilution	Xylose Concentration (μmol mL^{-1})
1:1 (undiluted)	10.0
1:2	5.0
1:3	3.3
1:5	2.0

Tubes containing the different standard solutions of xylose, with different concentrations, are submitted to the same procedure as the enzyme blank, as follows:

1)   750 µL of xylan 1 % (w/v);

2)   1.5 mL of DNS;

3)   250 µL of xylose solution;

4)   Boil for 5 min and cool;

5)   Measure the color produced at 540 nm against reagent blank.

As a result, different color intensities are generated by the different xylose concentrations, as shown in Fig. 5, and the standard curve is constructed using the different values of xylose concentration (Y axis) vs. the correspondent absorbance (X axis), Fig. 6.

Figure 5: Xylose standard showing the variation of colors generated by the different concentrations of xylose.

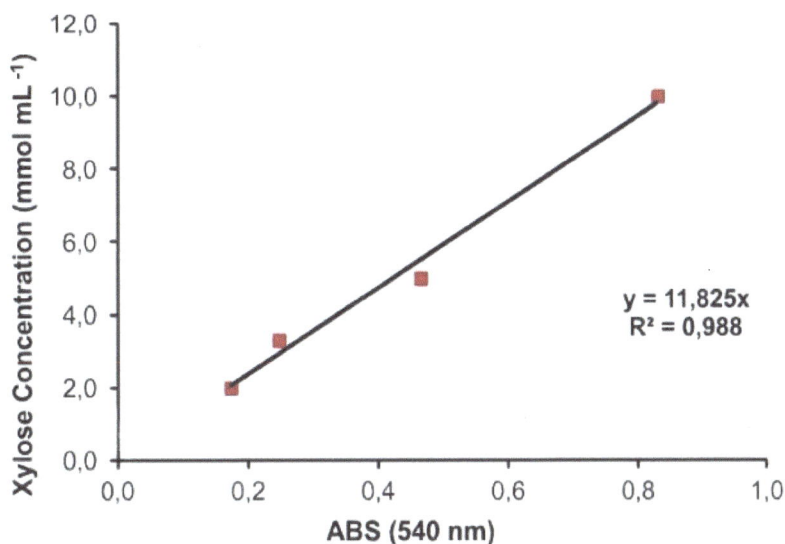

Figure 6: Xylose standard curve indicating the angular coefficient of the standard linear regression.

It is interesting to note that the standard curve can be prepared using concentration units (molarity or μmol mL-1), as shown in Fig. **6**, or amount units (micromoles or micrograms, for instance). If the standard curve is prepared using micromoles units, the calculation of the enzymatic activity is as specified in equation 1. However, if it is used any concentration or any amount unit different from micromoles, the calculation of the original activity must take this in account, and the proper transformations must be carried out.

Obs.: The standard curve should preferably be carried out using triplicates.

Comments

The procedure here presented for measurement of xylanase activity was adapted from Bailey *et al.* [11]. The variations made in the original methodology are concerning the use of minor volumes of solutions, which does not cause any interference in the results. For detecting reducing sugars, the method of DNS is more reliable, when compared to Nelson-Somogyi (NS). According to Grabskit and Jeffries [37] this is due to the fact that DNS reacts more intensely with the xylooligosaccharides present in the reaction mixture than with xylose. In the NS method, on the contrary, xylose is the main reacting sugar. The prevalence of oligosaccharides among the products of xylan hydrolysis leads to an underestimation of enzyme activity in NS method. Despite having a lower sensitivity than NS (10 times less), DNS technique is more widely used: the assay has less number of steps and was reproduced in various laboratories, presenting a standard deviation of 17 %, when a comparison between laboratories was made [38].

An important point to be stressed is that frequently it is necessary to make a dilution of the crude enzymatic extract for an appropriate assay of xylanase activity. Pay attention to the absorbance, obviously the maximal values obtained must be within the range of the standard curve. The standard curve, and also the enzyme and reagent blanks, must be performed in the presence of the substrate, xylan, since it may be colored and/or its insolubility may affect the readings in the spectrophotometer.

Several other methods for measuring xylanase activity are described in literature, but these are not used very often. For instance, the one that uses the chromogenic substrates RBB dye linked to xylan [11, 38, 39]. The use of the artificial chromogenic substrate p-nitrophenyl-β-D-xylobioside has also been cited [40-43], although studies about the use of this substrate are still needed. This type of compound is frequently used to test substrate affinity, and according to Puchart and Biely [38] it has a potential to fit as a preliminary phase for classification, to separate family GH10 from GH11 of xylanolytic enzymes.

In the same way, the fluorogenic substrate 4-methylumbelliferyl-β-D-xylobioside (MUX$_2$) has been applied [44] to detect and measure enzyme activities, providing superior sensitivity compared to chromogenic substrates. Due to higher sensitivity, small quantities of hydrolyzed substrate can be easily detected, allowing assays to be carried out at low substrate concentrations [45].

Measurement of β-xylosidase Activity Using a Chromogenic Compound, p-Nitrophenyl-β-D-Xylopyranoside

The complete degradation of xylan requires the presence of different enzymes. β-xylosidases acts on short-chain xylooligosaccharides and xylobiose, generating xylose [46, 47]. To evaluate the production of β-xylosidase the chromogenic substrate   p-nitrophenyl-β-D-xylopyranoside   (p-nitrophenyl-β-D-xyloside)   is frequently used.

Objective

To measure the enzymatic activity of β-xylosidase using a chromogenic substrate, p-nitrophenyl-β-D-xylopyranoside.

Principle

β-Xylosidase acts catalyzing the cleavage of the artificial substrate p-nitrophenyl-β-D-xylopyranoside liberating p-nitrophenol, which is likely to be measured [48].

Material and Equipment

1. Enzyme sample, *e.g.,* as crude extract or supernatant;

2. p-nitrophenyl-β-D-xylopyranoside 2.0 mM;

3. Acetate buffer 100 mM, pH 4.2;

4. Sodium carbonate solution 0.13 M;

5. Water bath with temperature controller;

6. Ice bath.

Procedure

Assay [49]

1. Add 0.5 mL enzyme sample (properly diluted, if necessary, in phosphate citrate buffer 50 mM pH 5.0) to 0.5 mL of p-nitrophenyl-β-D-xylopyranoside 2.0 mM in acetate buffer 100 mM pH 4.2;

2. Incubate the reaction mixture at 40 °C for 30 min;

3. Immerse on ice bath;

4. Stop the reaction using 3 mL of sodium carbonate solution 0.13 M;

5. Measure the color produced at 400 nm against the reagent blank [4];

6. Using the standard curve[5], convert the corrected absorbance to enzyme activity units (IU);

7. Calculate activity in the original (undiluted) sample by the following equation:

$$IU / mL^{1-} = \frac{ABS_{RM} \cdot \alpha \cdot D}{\Delta T} \qquad (2)$$

α= angular coefficient of the standard linear regression, in μmol. mL^{1-}

ABS $_{RM}$ = mean of absorbance of the reaction mixture

D= enzymatic dilution

ΔT= time of incubation (enzyme-substrate reaction), in minutes

IU= β-xylosidase activity is defined as the amount of enzyme required to produce 1 μmol of p-nitrophenol per minute of reaction

(4) *Reagent blank:*

✓ 0.5 mL of p-nitrophenyl-β-D-xylopyranoside 2.0 mM;

✓ 0.5 mL of acetate buffer 100 mM pH 4.2;

✓ Incubate 30 min, 40°C;

✓ 3 mL of sodium carbonate solution 0.13 M;

✓ Measure the color produced at 400 nm.

Obs.: This test should preferably be carried out using triplicates.

(5) *p-nitrophenol standard curve*

The standard used in this assay is p-nitrophenol (e.g., Sigma-Aldrich), prepared as stock solution (0.1 μmol mL^{-1}) in acetate buffer 100 mM pH 4.2. The stock solution can be diluted according to Tab. 2, however, if necessary, more dilutions can be added.

Table 2: Dilutions and respective p-nitrophenol concentration used in standard curve

Dilution	p-nitrophenol concentration (μmol mL^{-1})
1:1 (undiluted)	0.100
1:2	0.050
1:3	0.033
1:4	0.025
1:5	0.020

Tubes containing the different standard solutions of p-nitrophenol, with different concentrations, are submitted to the following procedure:

✓   0.5 mL of p-nitrophenol;

✓   0.5 mL of acetate buffer 100 mM pH 4.2;

✓   3 mL of sodium carbonate solution 0.13 M;

✓   Measure the color produced at 400 nm.

As a result different colors are generated by the different p-nitrophenol concentrations. The standard curve is constructed using the different values of each p-nitrophenol concentration (Y axis) *vs.* the correspondent absorbance (X axis).

It is interesting to note that the standard curve can be prepared using concentration units (molarity or μmol mL^{-1}, as shown in Fig. **6**) or amount units (micromoles or micrograms, for instance). If the standard curve is prepared using micromoles units the calculation of the enzymatic activity is as specified in equation 1. However, if it is used any concentration or any amount unit different from micromoles, the calculation of the original activity must take this in account, and the proper transformations must be carried out.

Obs.: The standard curve should preferably be carried out using triplicates.

Comments

Although there are some other methods to measure β-xylosidase activity, the technique here presented is the most common one cited in literature [50, 51]. There are variations concerning the concentration of substrate, time of incubation (substrate-enzyme reaction) and to the wavelength used to measure absorbance, which is generally around 400 nm.

Enzyme Characterization

Effect of Temperature and pH

The pH and temperature are important variables that interfere in enzymatic activity. Each enzyme has its own ideal pH and temperature (or range) which will give the ideal enzyme conformation to reach the best velocity of the enzymatic reaction. An ideal pH will also affect the ionizable substrates, which will allow substrate-enzyme interaction [52]. To evaluate the influence of those variables in xylanolytic activity, two different methodologies can be reported: classic methodology and experimental design. The first one uses a step by step system, where each variable is tested per time. The second option, which has been earning more and more users in the last 20 years, is based in a statistical technique, where both variables are tested at the same time, and possible interactions between them can be detected.

Objective

To determine the most adequate pH and temperature to be used in the reaction system (enzyme+substrate) for best enzymatic activity, under the studied conditions, using the classical procedure of step by step system.

Principle

The use of different values of pH and temperature to determine enzymatic activity will allow the detection of the best conditions for maximum activity.

Material and Equipment

1. Substrate;

2. Enzyme sample, *e.g.,* as crude extract or supernatant;

3. Buffers. *E.g.*: Glycine-NaOH 50 mM (pH 3.0); Citrate 50 mM (pH 4.0; 5.0; e 6.0); Citrate-phosphate (pH 7.0); Phosphate 50 mM (pH 8.0); Tris-HCl 50 mM (pH 9.0 e 10.0);

4. Water bath with temperature controller;

5. Ice bath.

Procedure

1. Choose a range of study of pH and temperature. *E.g.,* pH 3.0 to 10.0 and temperature 30-100 °C [53];

2. Prepare the substrate using different buffers;

3. Choose a fix pH. Measure the xylanolytic activity incubating the reaction mixture in a water bath with the fix pH and varying the incubation temperature, *e.g.,* Table **3**, and find the best temperature;

Table 3: Example of classic methodology showing the use of different temperatures in enzymatic reaction with a fix pH in order to achieve an ideal temperature\

pH^5	Temperature (^{o}C)
	30
	40
	50
	60
	70
	80
	90
	100

4. Measure the xylanolytic activity incubating the reaction mixture (substrate + enzyme) in a water bath with the best temperature found above and varying pH values, using substrates prepared with different buffers, Table **4**, so choosing the ideal pH.

Table 4: Example of classic methodology showing the use of different pH values in enzymatic reaction combined with best temperature in order to achieve the best pH

Temperature (^{o}C)	pH
	3
	4
Best temperature	5
	6
	7
	8
	9
	10

Obs.: This test should preferably be carried out using triplicates.

Comments

Tab. 5 illustrates some results of optima pH and temperature found in literature for xylanases produced by bacteria, actinomycetes and fungi. Values were, in a general way, in pH range from 5.0 to 10.0 and temperature from 40 to 80 °C.

Although we have presented the step by step technique for determining optima pH and temperature for xylanolytic activity, it must be emphasized that the use of experimental design results in fewer runs with a greater range of study. This is a very promising technique [53], however requires a previous knowledge of the statistical method for an adequate preparation of the experiments and interpretation of the results.

Table 5: Optimal pH and temperature of xylanases produced by different microorganism

Microorganism	Optimal pH	Optimal Temperature (°C)	Enzyme State	Refs.
Streptomyces fradiae var. 11	7.8	50	Purified enzyme	[54]
Bacillus amyloliquefaciens	6.8-7.0	80	Purified enzyme	[55]
Thermomyces lanuginosus NK-2	5.0	60	Partially purified	[56]
Bacillus subtillis	6.0	60	Purified enzyme	[57]
Aspergillus ficuum AF-98	5.0	45	Purified enzyme	[58]
Thermomyces lanuginosus CBS 288.54	7.0-7.5	70-75	Purified enzyme	[59]
Paecilomyces themophila	7.0	75-80	Purified enzyme	[60]
Penicillium expansum	5.5	40	Partially purified	[61]
Bacillus halodurans S7	9.0-9.5	75	Purified enzyme	[62]
Aspergillus fumigatus AR1	6.0-6.5	60-65	Crude enzyme	[63]
Burkholderia sp. DMAX	8.0-8.6	50	Partially purified	[64]
Streptomyces sp. Ab106	6.0	60	Crude enzyme	[65]
Streptomyces viridosporus T7A	7.0-8.0	65-70	Purified enzyme	[66]

Effect of Metal Ions

The enzymatic activity can be influenced by the presence of metal ions in the reaction mixture. Ions can act enhancing or inhibiting the enzymatic reactions, therefore, it is important to study their influence on enzyme activity.

Objective

To evaluate the effect of different metal ions on xylanolytic activity.

Principle

This technique is based in a comparison between the enzyme-substrate reaction system with and without the addition of metal ion in order to see its effect on the reaction.

Material and Equipment

1. Enzyme sample, *e.g.,* as crude extract or supernatant;

2. Metal ion solution;

3. Substrate;

4. Water bath with temperature controller;

5. Ice bath.

Procedure

1. Measure xylanolytic activity in the presence and in the absence of the metal ion solution of interest;

2. Calculate the relative activity of the reaction (%) using the metal ion.

Comments

Almost one third of all known enzymes require one or more metal ions for catalytic activity acting in their conformation. Enzymes can be modulated by interaction of cations with amino acid residues involved in their active sites. Such interactions can either increase (positive modulation) or diminish (negative modulation) the enzyme catalytic activity [67]. Table **6** shows examples of different metal ions and their influence on xylanolytic activity. As can be seen, xylanolytic activity from different microorganisms is differently influenced by metal ions. Ion Hg^{2+} appears as a great inhibitor, as in three of the six articles described, its presence resulted in no activity.

It is interesting to stress that other compounds besides metallic ions can be tested for inhibition or enhancement of xylanolytic activity, using the same protocol.

This includes any compound as for instance EDTA, SDS, DMSO, β-mercaptoethanol [68]. Also different ion concentrations must be used, depending on the purpose of the experiments.

Table 6: Influence of metal ions on xylanolytic activity

	Effect of Metal Ions (relative activity %)							
	Microorganisms							
	Bacillus halodurans S7 [62]		*Streptomyces malaysiensis* AMT-3 [13]	*Bacillus circulans* BL53 [68]	*Streptomyces fradiae* var. k11 [54]	*Aspergillus fumigatus* AR1 [63]	*Bacillus amyloliquefaciens* [55]	
Metal Ion \ Molarity	1 mM	5 mM	10 mM	2 mM	1 mM	5 mM	1 mM	5 mM
Mn^{2+}	36	3	17.7	161.88	110.7	117.2	134.3	106.1
Cu^{2+}	107	13	18.3	91.44	87.9	133.0	6.3	5.9
Co^{2+}	119	63	90.5	151.67	109.6	-	-	-
Ca^{2+}	103	81	92.1	95.39	90.7	84.0	101.4	91.5
Fe^{2+}	110	103	16.0	81.90	161.9	73.7	78.4	70.8
Zn^{+2}	86	73	88.5	91.11	96.6	-	103.8	99.3
Sn^{2+}	109	125	-	-	-	-	-	-
Ni^{2+}	107	122	-	-	102.0	102.0	-	-
Ru^{2+}	95	105	-	-	-	-	-	-
K^+	100	103	92.7	-	92.5	95.0	-	-
Mg^{+2}	104	90	91.3	82.55	98.7	98.0	87.6	102.9
Na^+	102	101	107.5	92.10	87.4	104.7	110.6	93.8
Hg^{+2}	90	75	-	0.00	0.0	68.5	37.8	0
Fe^{3+}	-	-	-	-	100.3	-	57.1	6.7
Pb^{2+}	-	-	-	78.93	99.5	-	-	-
Li^+	-	-	-	-	101.9	-	94.5	93.2
Cr^{3+}	-	-	-	-	102.7	-	-	-
Ag^+	-	-	-	-	-	109.0	-	-

Zymogram

Zymogram is a useful tool to study the xylanases profiles produced by microorganisms. This technique provides the visualization of the enzymatic activity, due to the utilization of the substrate (xylan) copolymerized with the polyacrylamide gel.

The detection of xylanolytic activity can be performed by the use of fluorogenic substrate, like methylumbelliferyl [69], the use of a chromogenic substrate, like RBB linked to xylan [70, 71, 72] or by the use of natural substrate revealed by the addition of Congo Red solution, as described below.

Objective

Study the profile of xylanases using a sodium-dodecil-sulphate polyacrylamide gel electrophoresis (SDS-PAGE) system containing xylan, and determine the apparent molecular masses of each of the enzymes.

Principle

After a run in an electrophoresis device, where xylanases are separated according to its molecular masses, the enzymes are allowed to cleave the substrate, xylan, present in the polyacrylamide gel, producing hydrolysis bands which can be visualized by the addition of Congo Red.

Material and Equipment

1. Stacking gel;

2. Running gel;

3. Enzyme sample, *e.g.,* as crude extract or supernatant;

4. Molecular mass marker (*i.e.,* 12 – 225 kDa - Amersham);

5. Electrophoresis system;

6. Triton X-100 (1 % v/v in water), kept under refrigeration;

7. Orbital shaker;

8. Incubation buffer (appropriated pH);

9. Water bath adjusted to appropriated temperature;

10. Congo Red (0.1 % w/v);

11. 1 M NaCl;

12. Distilled water;

13. Transilluminator with visible light;

14. Running buffer;

15. Sample buffer.

Procedure

Preparing the Stacking Gel

Solutions	Concentration	Volume (μl)
Distilled water	-	1700
Bis-acrylamide [6]	30 % (w/v)	500
Tris-HCl buffer (pH 6.8) [7]	0.5 M	750
Sodium dodecil sulphate (SDS)	10 % (w/v)	30
Amonium persulphate (APS)	10 % (w/v)	30
Tetramethylethylenediamine (TEMED)	-	3

(6) *Bis-acrylamide - Acrylamide (60 g) + Bis-Acrylamide (1.6 g). Complete to 200 mL with distilled water. Stock at 4 °C away from light.*

(7) *Tris-HCl buffer 0.5 M (pH 6.8) - Tris (6 g) + HCl 4 N (until pH 6.8). Complete to 200 mL with distilled water.*

Preparing the Running Gel

Solutions	Concentration	Volume (μl)
Distilled water	-	2300.0
Bis-acrylamide	30 % (w/v)	3000.0
Tris-HCl buffer (pH 8.8) [8]	1.5 M	2250.0
Oat spelts xylan	0.1 % (w/v)	1250.0

SDS	10 % (w/v)	125.0
APS	10 % (w/v)	12.5
TEMED	-	12.5

(8)　*Tris-HCl 1.5 M (pH 8.8) – Tris (36,3g) + HCl 4 N (until pH 8.8). Complete to 200 mL with distilled water.*

Preparing the Running Buffer

Solutions	Amount
SDS	1.0 g
Glycine	14.4 g
Tris	3.03 g
Distilled water	q.s. 1000 mL

Obs.: The running buffer must be kept under refrigeration.

Preparing the Sample Buffer

Solutions	Amount
SDS 10% (w/v)	4.0 mL
Glycerol	2.0 mL
Tris-HCl 0.5 M pH 6.8	2.5 mL
β-mercaptoethanol 2 % v/v or Dithiotreitol (DTT)	0.2 mL/ 0.31 g
Bromophenol Blue 0.02 % (w/v)	0.2 mg
Distilled water	q.s. 10 mL

Obs.: The sample buffer must be kept in freezer at -20oC.

Electrophoresis

1. Prepare running gel: a polyacrylamide gel (usually 10 % w/v) containing oat spelts xylan 0.1 % (w/v) (or other type of xylan);

2. Add tetramethylethylenediamine (TEMED) to catalyze acrylamide polymerization in the running gel;

3. Wait 50 min and check polymerization;

4. Prepare stacking gel: a polyacrylamide gel 3 % (w/v);

5. Add TEMED to stacking gel and accommodate the stacking gel above the running gel;

6. Insert the comb into the stacking gel to generate the wells;

7. Wait 30-40 min and check polymerization;

8. Remove the comb and release the gel base;

9. Place the gel into the electrophoresis cube and leave it at 4 °C (until the beginning of the run);

10. Add the running buffer covering the whole system;

11. Add 50 µL (or other volume according to the size of the comb and gel) of the sample containing enzyme plus sample buffer (the exact volume of the enzyme will depend on the enzyme activity and/or protein content);

12. Add around 6 µL of molecular mass marker (*e.g.,* 12-225 kDa – pre-stained Amersham);

13. Close the system and run the gel at 100 V and 30 mA until bromophenol blue, present in the sample buffer, almost reaches the end of the gel (the system must be maintained under refrigeration;

14. Transfer the gel to a recipient containing the renaturing solution (Triton X-100 1 % v/v);

15. Keep the recipient away from light and incubate in ice bath for 60 min, under agitation;

16. Wash the gel with distilled water to remove the renaturing solution;

17. Incubate the gel in the ideal pH buffer and temperature during an appropriate time (*e.g.,* 20 min);

18. Wash the gel with distilled water [9];

19. Submerge in 0.1 % (w/v) Congo Red solution for 10 min in orbital agitation [10];

20. Wash the gel with 1 M NaCl (successively) until visualization of enzyme bands;

21. With the aid of a transiluminator, observe xylanolytic bands indicating enzymatic activity (Fig. **7**);

22. Calculate molecular masses [11].

(9)   If a buffer of low pH is used in incubation period, the gel might become blue, as explained before on Congo Red's qualitative assay for xylanase activity. For this reason, it is necessary washing the gel after incubation.

(10) When using a pre-stained standard molecular masses the position of the migration of each protein can disappear after Congo Red staining. For this reason you may mark these positions in the gel before staining.

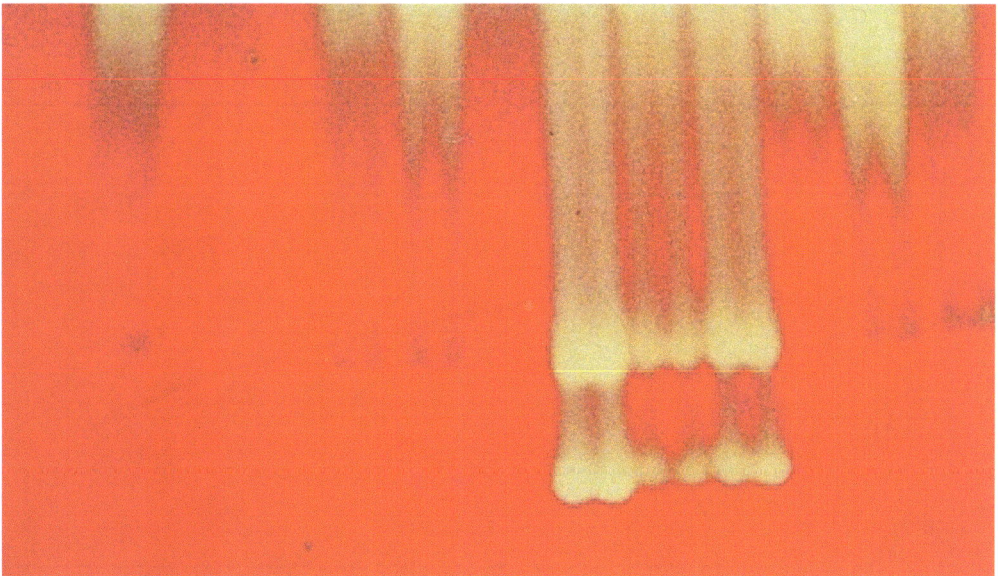

Figure 7: Zymogram showing the decolorized xylanolytic bands.

(11) Calculating the apparent molecular masses of xylanases:

The use of molecular mass markers make possible the deduction of an apparent molecular mass for xylanases forms visualized on electrophoresis gel after revelation using Congo Red. To achieve this, a standard line using a logarithmic curve must be built. The curve shows a correlation between molecular masses and the distance measured in centimeters. This corresponds to the running of the molecular mass marker in gel from the well where it was applied. Table 7 shows an example using molecular mass marker and the distance they ran in an electrophoresis gel.

Table 7: Standard curve: example of molecular masses from standards proteins and the distances they migrated in an electrophoresis gel.

Molecular masses marker (kDa)	Distance (cm)
225	0.7
150	1
102	1.5
76	2
52	3.1
38	4.1
31	4.8
24	5.5

Using the formula of the standard curve, obtained from the standards of molecular masses, Fig. **8**, the apparent molecular masses of xylanase forms found in gel can be deduced. The distance between the band and the beginning of the running gel is used as the independent variable of the formula, resulting in an apparent molecular mass of the band.

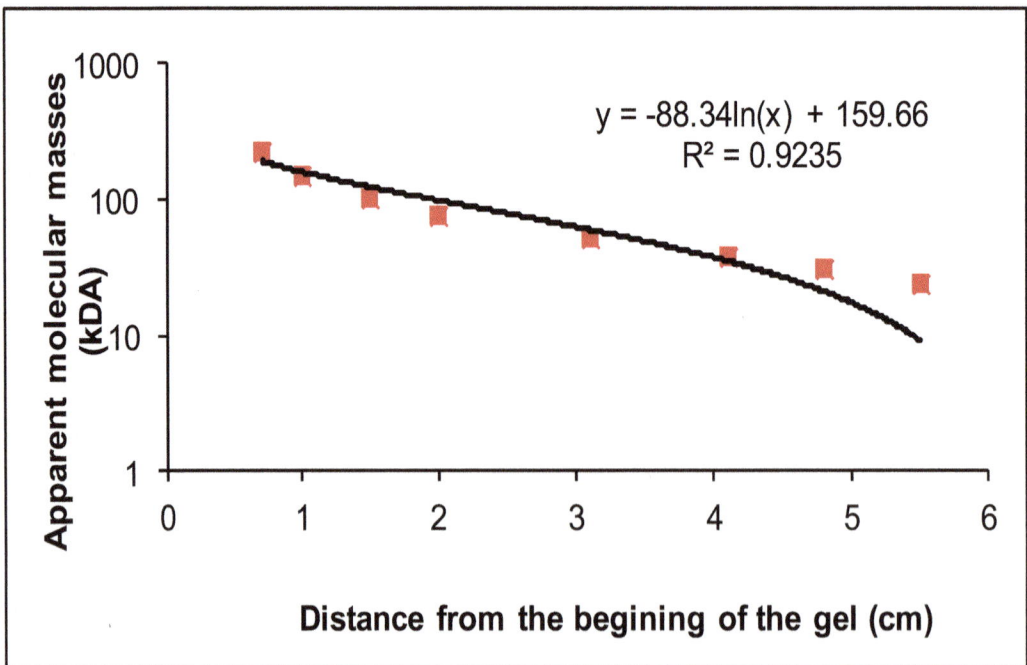

$$y = -88.34\ln(x) + 159.66$$
$$R^2 = 0.9235$$

Figure 8: Standard curve using apparent molecular mass of standards related to their migration in electrophoresis gel.

Comments

Different types of xylan can be used, such as birchwood, larchwood, *etc.* The dye Congo Red exhibits high affinity for polysaccharides like xylan, so hydrolysis zones can be visualized in contrast with the color generated by the dye. It is also possible to estimate the molecular masses of xylanases obtained from samples by comparing them with the standard molecular weight.

PERSPECTIVES

The last 20 years have seen a growing interest in microbial enzyme systems that degrade hemicelluloses, in special xylan. This chapter presented a range of qualitative and quantitative approaches for the assessment of xylan degrading microorganisms and xylanase activity. The choice of assay method and conditions under which the tests are to be taken will depend, among other factors, on growth characteristics (for screening), availability of reagents and enzyme physicochemical characteristics. Considering the enzymatic activity assay according to Bailey *et al.* [11], xylan concentration is 1 % (w/v) and reaction time is 30 min, however these conditions can be optimized. By determining some kinetic parameters of the enzymatic reaction, such as Michaelis constant (K_m) and maximal velocity (V_m), the best substrate concentration and the best time reaction can be found [67].

It is worthwhile to point out the great advantage of using the experimental design techniques for optimizing pH and temperatures for the xylanase enzymatic assay. For instance, using the step by step technique, to study 5 different pH and 5 temperatures would culminate in 10 runs with a limitation of results due to the fixation of one variable. On the other hand, to study all possible variations of pH and temperature it would be necessary the use of 25 runs. Using an experimental design, in this case, specifically, the central composed rotational design (CCRD), with only 11 runs, it is possible to obtain results including not only the best conditions but also the possible interference of one variable with another. Although these techniques does not seem very simple in a first sight, they are very useful, being economical and less time consuming.

Also important to emphasize, is that the lack of standardization of the techniques used to measure xylanase activity, it impairs an effective comparison between the results presented by the various researchers. The use of different substrates may cause variable results, due to different xylanase affinities to the substrate and also due to different reagent´s affinity to the various compounds generated by xylanase action. It would be of great interest if laboratories all over the world could discuss the existing methodologies, looking not only for standardization, but also to the economical aspects concerning substrates and reagents as a whole.

ACKNOWLEDGEMENTS

This work was supported by Fundação Carlos Chagas Filho de Amparo a Pesquisa do Estado do Rio de Janeiro (FAPERJ), Conselho Nacional de Desenvolvimento Científico e Tecnológico (MCT/CNPq) and Coordenação de Aperfeiçoamento de Pessoal do Ensino Superior (CAPES).

CONFLICT OF INTEREST

The authors confirm that this chapter content has no conflict of interest.

REFERENCES

[1]   PRADE, R. A. Xylanases: from biology to biotechnology. Biotech Genet Eng Rev 1995; 13: 100–31.
[2]   COLLINS, T.; GERDAY, C.; FELLER, G. Xylanases, xylanase families and extremophilic xylanases. FEMS Microbiol Rev 2005; 29: 3-23.
[3]   SUNNA, A.; ANTRANIKIAN, G. Xylanolytic enzymes from fungi and bacteria. Crit Rev Biotechnol 1997; 17: 39-67.
[4]   UHL, A. M.; DANIEL, R. M. The first description of an archaeal hemicellulase: the xylanase from *Thermococcus zilligii* strain AN1. Extremophiles 1999; 3: 263–7.
[5]   LUTHI, E.; JASMAT, N. B.; BERGQUIST, P. L. Xylanase from the extremely thermophilic bacterium *Caldocellum saccharolyticum*: overexpression of the gene in *Escherichia coli* and characterization of the gene product. Appl Environ Microbiol 1990; 56: 2677-83.
[6]   KHASIN, A.; ALCHANATI, I.; SHOHAM, Y. Purification and characterization of a thermostable xylanase from *Bacillus stearothermophilus* T-6. Appl Environ Microbiol 1993; 59: 1725–30.
[7]   LO LEGGIO, L.; KALOGIANNIS, S.; BHAT, M. K. *et al*. High resolution structure and sequence of *T. aurantiacus* xylanase I: implications for the evolution of thermostability in family 10 xylanases and enzymes with (beta) alpha-barrel architecture. Proteins 1999; 36: 295–306.

[8]   ZVERLOV, V.; PIOTUKH, K.; DAKHOVA, O. *et al*. The multidomain xylanase A of the hyperthermophilic bacterium *Thermotoga neapolitana* is extremely thermoresistant. Appl Microbiol Biotechnol 1996; 45: 245–7.

[9]   ABOU-HACHEM, M.; OLSSON, F.; KARLSSON, E. N. Probing the stability of the modular family 10 xylanase from *Rhodothermus marinus*. Extremophiles 2003; 7: 483–91.

[10]   SHALLOM, D.; SHOHAM, Y. Microbial hemicellulases. Curr Opin Microbiol 2003; 6: 219–28.

[11]   BAILEY, M. J.; BIELY, P.; POUTANEN, K. Interlaboratory testing of methods for assay of xylanase activity. J Biotechnol 1992; 23: 257-70.

[12]   NASCIMENTO, R. P.; MARQUES, S.; ALVES, L. *et al*. A novel strain of *Streptomyces malaysiensis* isolated from Brazilian soil produces high endo-β-1,4-endoxylanase titres. J Microbiol Biotechnol 2003; 19: 879-81.

[13]   SALEEM, F.; AHMED, S.; JAMIL, A. Isolation of a xylan degrading gene from genomic DNA library of a thermophilic fungus *Chaetomium thermophile* ATCC 28076. Pak J Bot 2008; 40(3): 1225-30.

[14]   MENG, X.; ZONGZE, S.; YUZHI, H. *et al*. A novel pH-stable, bifunctional xylanase isolated from a deep-Sea microorganism, *Demequina* sp. JK4. J Ind Microbiol Biotechnol 2009; 19(10): 1077–84.

[15]   HOU, Y. H.; WANG, T. H.; LONG, H. *et al*. Novel cold-adaptive *Penicillium* strain FS010 secreting thermo-labile xylanase isolated from Yellow sea. Acta Biochim Biophys Sin 2006; 38(2): 142–9.

[16]   LEE, T. H.; PYUNG, O. L.; YONG-EOK, L. Cloning, characterization, and expression of xylanase A gene from *Paenibacillus* sp. DG-22 in *Escherichia coli*. J Microbiol Biotechnol 2007; 17(1): 29–36.

[17]   SQUINA, F. M.; MORT, A. J.; DECKER, S. R. *et al*. Xylan decomposition by *Aspergillus clavatus* endo-xylanase. Protein Expr Purif 2009; 68: 65–71.

[18]   TEATHER, R. M.; WOOD, P. J. Use of Congo Red-polysaccharide interactions in enumeration and characterization of cellulolytic bacteria from the bovine rumen. Appl Environ Microbiol 1982; 56: 777-80.

[19]   BRECCIA, J. D.; CASTRO, G. R.; BAIGORI, M. D. *et al*. Screening of xylanolytic bacteria using a colour plate method. J Appl Bacteriol 1995; 78: 469-72.

[20]   RAGHUKUMAR, C.; MURALEEDHARAN, U.; GAUD, V. R.; *et al*. Xylanases of marine fungi of potential use for biobleaching of paper pulp. J Ind Microbiol Biotechnol 2004; 31(9): 433-41.

[21]   SHARMA, P.; PAJNI, S.; DHILLON, N. *et al*. Limitations of Congo Red staining techniques for the detection of cellulolytic activity. Biotechnol Lett 1986; 8(8): 579-80.

[22]   STROBAND, H. W. J; KROON, A. G. The development of the stomach in *Clarias lazera* and the intestinal absorption of protein macromolecules. Cell Tissue Res 1981; 215: 397-415.

[23]   KHURANA, R.; UVERSKY, V. N.; NIELSEN, L.; FINK, A. L. Is Congo Red an amyloid-specific dye? J Biol Chem 2001; 276(25): 22715–21.

[24]   TEN, L. N.; IM, W. T.; KIM, M. K. *et al*. Development of a plate technique for screening of polysaccharide-degrading microorganisms by using a mixture of insoluble chromogenic substrates. J Microbiol Methods 2004; 56: 375-82.

[25]   KIISKINEN, M. R.; RÄTTO, M.; KRUUS, K. Screening for novel laccase-producing microbes. J Appl Microbiol 2004; 97: 640–64.

[26]   AKPAN, I.; BANKOLE, M. O.; ADESEMOWO, A. M. A rapid plate culture method for screening of amylase producing micro-organisms. Biotechnol Tech 1999; 13(6): 411-13.

[27]   FARKAS, V.; LISKOVÁ, M.; BIELY, P. Novel media for detection of microbial producers of cellulase and xylanase. FEMS Microbiol Lett 1985; 28: 137-40.

[28]   LEE, H.; BIELY, P.; LATTA, R. K. *et al.* Utilization of xylan by yeasts and its conversion to ethanol by *Pichia stipitis* strains. Appl Environ Microbiol 1986; 52(2): 320-4.

[29]   WHITEHEAD, T. R.; HESPELL, R. B. The genes for three xylan-degrading activities from *Bacteroides ovatus* are clustered in a 3.8-kilobase region. J Bacteriol 1990; 172(5): 2408-12.

[30]   GILBERT, H. J.; SULLIVAN, D.A.; JENKINS, G. *et al.* Molecular cloning of multiple xylanase genes from *Pseudomonas fluorescens* subsp. cellulose. J Gen Microbiol, 1988; 134: 3239-47.

[31]   RAHA, A. R.; CHANG, L. Y.; SIPAT, A. *et al.* Expression of a thermostable xylanase gene from *Bacillus coagulans* ST-6 in *Lactococcus lacti.* Lett Appl Microbiol 2006; 42(3): 210–4.

[32]   STRAUSS, M. L. A.; JOLLY, N. P.; LAMBRECHTS, M. G. *et al.* Screening for the production of extracellular hydrolytic enzymes by non-*Saccharomyces* wine yeasts. J Appl Microbiol 2001; 91: 182-90.

[33]   SILVA, C. H. C.; PULS, J.; SOUSA, M. V. *et al.* Purification and characterization of a low molecular weight xylanase from solid-state cultures of *Aspergillus fumigatus* fresenius. Rev Microbiol 1999; 30: 114-9.

[34]   FAULET, B. M.; NIAMKÉ, S.; GONNETY, J. T. *et al.* Purification and biochemical properties of a new thermostable xylanase from symbiotic fungus, *Termitomyces* sp. African J Biotechnol 2006; 5(3): 273-82.

[35]   PUCHART, V.; BIELY, P. A simple enzymatic synthesis of 4-nitrophenyl β-1,4-D-xylobioside, a chromogenic substrate for assay and differentiation of endoxylanases. J Biotechnol 2007; 128: 576–86.

[36]   MILLER, G. L. Use of dinitrosalicylic acid reagent for determination of reducing sugar. Anal Chem 1959; 31(3): 426–28.

[37]   GRABSKIT, A. C.; JEFFRIES, T. W. Production, purification, and characterization of β-(1-4)-endoxylanase of *Streptomyces roseiscleroticus.* Appl Environ Microbiol 1991; 57(4): 987-92.

[38]   BIELY, P.; PUCHART, V. Recent progress in the assays of xylanolytic enzymes. J Sci Food Agric 2006; 86: 1636–47.

[39]   SEWELL, G. W.; ALDRICH, H. C.; WILLIANS, D. *et al.* Isolation and characterization of xylan-degrading strains of *Butyrivibrio fibrisolvens* from a napier grass-fed anaerobic digester. Appl Environ Microbiol 1988; 54(5): 1085-90.

[40]   TAGUCHI, H.; HAMASAKI, T.; AKAMATSU, T. *et al.* A simple assay for xylanase using o-nitrophenil-β-D-xylobioside. Biosc Biotech Biochem 1996; 60(6): 983-5.

[41]   KUNO, A.; SHIMIZU, D.; KANEKO, S.*et al.* Significant enhancement in the binding of p-nitrophenyl-β-D-xylobioside by the E128H mutant F/10 xylanase from *Streptomyces olivaceoviridis* E-86. FEBS Lett 1999; 450: 299-305.

[42]   HONDA, Y.; KITAOKA, M.; SAKKA, K. *et al.* An investigation of the pH-activity relationships of Cex, a family 10 xylanase from *Cellulomonas fimi:* Xylan inhibition and the influence of nitro-substituted aryl-β-D-xylobiosides on xylanase activity. J Biosci Bioeng 2002; 93(3): 313-7.

[43] KANEKO, S.; ICHINOSE, H.; FUJIMOTO, Z. *et al*. Structure and function of a family 10 β-xylanase chimera of *Streptomyces olivaceoviridis* E-86 FXYN and *Cellulomonas fimi* Cex. J Biol Chem 2004; 279(25): 26619–26.

[44] CHRISTAKOPOULOS, P.; NERINCKX, W.; KEKOS, D. *et al*. Purification and characterization of two low molecular mass alkaline xylanases from *Fusarium oxysporum* F3. J Biotechnol 1996; 51: 181-9.

[45] MARX, M. C.; WOOD, M.; JARVIS, S. C. A microplate fluorometric assay for the study of enzyme diversity in soils. Soil Biol Biochem 2001; 33: 1633-40.

[46] FREDERICK, M. M.; FREDERICK, J. R.; FRATZKE, A. R. *et al*. Purification and characterization of a xylobiose and xylose-producing endo-xylanase from *Aspergillus niger*. Carbohyd Res 1981; 97: 87-103.

[47] GÓMEZ, M.; ISORNA, P.; ROJO, M. *et al*. Kinect mechanism of β-xylosidase from *Trichoderma reesei* QM 9414. J Mol Catal B: Enzym 2001, 16: 7-15.

[48] COUGHAN, M. P.; HAZLEWOOD, G. P. B-1,4-D-xylan-degrading enzyme systems: biochemistry, molecular biology and applcations. Biotechnol Appl Biochem 1993; 17: 259-89.

[49] GHOSE, T. K.; BISARIA, V. S. Measurement of hemicellulase activities part 1: xylanases. Pure Appl Chem 1987; 59(12): 1739-52.

[50] MARGOLLES-CLARK, E.; TENKANEN, M.; NAKARI-SETALA, T. *et al*. Cloning of genes encoding a-L-arabinofuranosidase and β-xylosidase from *Trichoderma reesei* by expression in *Saccharomyces cerevisiae*. Appl Environ Microbiol 1996; 62(10): 3840–6.

[51] KADOWAKI, M. K.; SOUZA, C. G. M.; SIMÃO, R. C. G. *et al*. Xylanase production by *Aspergillus tamari*. Appl Biochem Biotechnol 1997; 66(2): 97-106.

[52] LIMA, U. A.; AQUARONE, E.; BORZANI, W.; SCHMIDELL, W. Eds. Elementos de Enzimologia: Biotecnologia Industrial. São Paulo: Edgard Blücher Ltda., 2001; pp. 165-71.

[53] RODRIGUES, M. I.; IEMMA, A. F. Planejamento de Experimentos e Otimização de Processos. Cárita. Campinas, Brazil, 2009.

[54] LI, N.; YANG, P.; WANG, Y. *et al*. Cloning, expression, and characterization of protease-resistant xylanase from *Streptomyces fradiae* var. k11. J Microbiol Biotechnol 2008; 18(3): 410–6.

[55] BRECCIA, J. D.; SIÑERIZ, F.; BAIGORÍ, M. D. *et al*. Purification and characterization of a thermostable xylanase from *Bacillus amyloliquefaciens*. Enzyme Microb Technol 1998; 22(1): 42-9.

[56] NAVEEN, K.; AGRAWAL, S. C.; JAIN, P. C. Production and properties of a thermostable xylanase by *Thermomyces lanuginosus* NK-2 grown on lignocelluloses. Biotechnology 2006; 5(2): 148-52.

[57] SÁ-PEREIRA, P.; MESQUITA, A.; DUARTE, J.C. *et al*. Rapid production of thermostable cellulase-free xylanase by a strain of *Bacillus subtilis* and its properties. Enzyme Microb Technol 2002; 30: 924–33.

[58] FENGXIA, L.; MEI, L.; ZHAOXIN, L. *et al*. Purification and characterization of xylanase from *Aspergillus ficuum* AF-98. Bioresour Technol 2008; 99: 5938–41.

[59] LI, X. T.; JIANG, Z. Q.; LI, L. T. *et al*. Characterization of a cellulase-free, neutral xylanase from *Thermomyces lanuginosus* CBS288.54 and its biobleaching effect on wheat straw pulp. Bioresour Technol 2005; 96: 1370–9.

[60] LI, L.; TIAN, H.; CHENG, Y. *et al*. Purification and characterization of a thermostable cellulase-free xylanase from the newly isolated *Paecilomyces themophila*. Enzyme Microb Technol 2006; 38: 780–7.

[61] QUERIDO, A. L. S.; COELHO, J. L. C.; ARAÚJO, E. F. *et al.* Partial purification and characterization of xylanase produced by *Penicillium expansum*. Braz Arch Biol Technol 2006; 49(3): 475-80.

[62] MAMO, G.; HATTI-KAUL, R.; MATTIASSON, B. A thermostable alkaline active endo-β-1-4-xylanase from *Bacillus halodurans* S7: Purification and characterization. Enzyme Microb Technol 2006; 39: 1492–8.

[63] ANTHONY, T.; RAJ, C. K.; RAJENDRAN, A. *et al.* High molecular weight cellulase-free xylanase from alkali-tolerant *Aspergillus fumigatus* AR1. Enzyme Microb Technol 2003; 32: 647–54.

[64] MOHANA, S.; SHAH, A.; DIVECHA, J. *et al.* Xylanase production by *Burkholderia* sp. DMAX strain under solid state fermentation using distillery spent wash. Bioresour Technol 2008; 99: 7553–64.

[65] TECHAPUN, C.; CHAROENROET, T.; POOSARAN, N. *et al.* Thermostable and alkaline-tolerant cellulase-free xylanase produced by thermotolerant *Streptomyces* sp. Ab106. J Biosci Bioeng 2002; 93(4): 431-3.

[66] MAGNUSON, T. S.; CRAWFORD, D. L. Purification and characterization of an alkaline xylanase from *Streptomyces viridosporus* T7A. Enzyme Microb Technol 1997; 21: 160-4.

[67] NELSON, D. L.; COX, M. M. Eds. Lehninger. Principles of Biochemistry. New York, W.H. Freeman and Company, 2008.

[68] HECK, J. X.; SOARES, L. H. B.; HERTZ, P.F. *et al.* Purification and properties of a xylanase produced by *Bacillus circulans* BL53 on solid-state cultivation. Biochem Eng J 2006; 32: 179–84.

[69] KALOGERIS, E.; CHRISTAKOPOULOS, P.; KEKOS, D.; MACRIS, B. J. Studies on the solid-state production of thermostable endoxylanases from *Thermoascus aurantiacus*: Characterization of two isozymes. J Biotechnol 1998; 60: 155–63.

[70] ROYER, J. C.; NAKAS, J. P. Simple, sensitive zymogram technique for detection of xylanase activity in polyacrylamide gels. Appl Environ Microbiol 1990; 56(6): 1516-7.

[71] REIS, S.; COSTA, M. A. F; PERALTA, R. M. Xylanase production by a wild strain of *Aspergillus nidulans*. Acta Sci Biol Sci 2003; 25(1): 221-5.

[72] GHATORA, S. K.; CHADHA, B. S.; BADHAN, A. K. *et al.* Identification and characterization of diverse xylanases from thermophilic and thermotolerant fungi. Bioresour 2006; 1(1): 18-33.

Send Orders for Reprints to reprints@benthamscience.net

Methods to Determine Enzymatic Activity, 2013, 161-194

CHAPTER 7

Assay Methods for Lipase Activity

Mônica Caramez Triches Damaso[1,*], Thaís Fabiana Chan Salum[1], Selma Da Costa Terzi[2] and Sonia Couri[3]

[1]Embrapa Agroenergy, Brasília, Brazil; [2]Embrapa Food Tecnology, Rio de Janeiro, Brazil and [3]Federal Institute of Education, Science and Technology of Rio de Janeiro, Rio de Janeiro, Brazil

Abstract: Lipases hydrolyze triacylglycerols to diacylglycerols, monoacylglycerols, fatty acids, and glycerol in their natural environment. However, in non-aqueous media, these enzymes catalyze reverse reactions such as esterification and transesterification. The versatility of the molecular structure and catalytic properties of these enzymes allow them to be used in various industries such as: food, cosmetic, oleochemical, pharmaceutical and detergent. Since lipases have several applications and have been the focus of numerous research studies, this chapter contains the guidelines for screening of microorganisms that produce lipases and for meticulous enzymatic analyses, focusing on the need to standardize methods in order to compare the catalytic efficiency of lipases obtained by different sources in distinct laboratories.

Keywords: Lipase, esterase, triacylglycerol, fatty acid, microorganism, titrimetry, spectroscopy, screening, enzymatic index, hydrolysis, esterification, synthesis, p-nitrofenil palmitate, olive oil, cupric acetate, rhodamine B, tributyrin, triolein, oleic acid.

INTRODUCTION

Lipases (glycerol ester hydrolases EC 3.1.1.3) are a family of enzymes that in their natural environment hydrolyze triacylglycerols to diacylglycerols, monoacylglycerols, fatty acids and glycerol. However, in non-aqueous media, lipases have shown to be very active in esterification and transesterification reactions (interesterification, alcoholysis, acidolysis, aminolysis) [1-3] (Fig. **1**).

The catalytic action of lipases is very complex and the enzyme structure around the active site varies from one to another lipase. Nevertheless, there are structural motifs common to all lipases: the catalytic triad, consisting of the amino acids

*Address correspondence to Mônica Caramez Triches Damaso: Embrapa Agroenergy, Brasilia, Brazil; Tel: +55 (61) 34482328; Fax: +55 (61) 34481589; E-mail: monica.damaso@embrapa.br.

serine, histidine and aspartate (or glutamate) [4] and the α/β hydrolase fold, a characteristic fold that is common to several hydrolytic enzymes [5].

Hydrolysis Reaction

Esterification Reaction

Transesterification Reactions

Alcoholyse

Acidolysis

Interesterification

Aminolysis

Figure 1: Reactions catalyzed by lipases.

A known phenomenon of lipolytic reactions is the fact that the activity of lipases is enhanced towards insoluble substrates that form an emulsion. This property became known as 'interfacial activation' and was used for a long time to distinguish lipases from esterases. The determination of lipases 3D structures seemed to provide an explanation for interfacial activation: most lipases have a short α-helix that covers their active sites, which is called lid or flap. Upon binding to a water-lipid interface, the lid moves away, and the closed lipase form turns into an open form, in which the active site is exposed. More recently, it was found that the presence of the lid is not necessarily related to interfacial activation: lipases from *Pseudomonas aeruginosa, Candida antarctica* B and *Burkholderia glumae* do not show interfacial activation even though they have the lid covering their active sites. Furthermore, some lipases, such as *Staphylococus hyicus*, show interfacial activation only with some substrates. Therefore, the presence of a lid and interfacial activation are unsuitable criteria to determine whether a specific esterase belongs to the lipase sub-family or not [6]. Then, as it was concluded that lipases have the capacity to act on solutions as well as emulsions, they should simply be defined as a particular class of carboxyl ester hydrolases having, unlike esterases, the capacity to act on water-insoluble long-chain acylglycerols [7].

Lipases constitute the most important group of biocatalysts for biotechnological applications. These enzymes are so attractive because of three aspects: 1) their chemoselectivity, regioselectivity and stereoselectivity properties; 2) the potential to be obtained in large amounts by microorganisms; and 3) crystal structures of several lipases have been solved, facilitating the design of genetic engineering strategies [8]. Lipases find promising applications in many industries such as oleochemical, food, cosmetic, pharmaceutical, agrochemical, detergent, biosurfactants and paper [1, 8].

Lipolytic enzymes such as lipases are indispensable for the bioconversion of lipids in the human organism and can be secreted by microorganisms [9-11], plants [12] and animals [13].

Microbial lipases are often more useful than those derived from plants or animals because of the rapid growth of microorganisms, ease of genetic manipulation and since they do not depend on seasonal issues [14]. Some examples of microorganisms that produce lipases are: *Aspergillus niger* [15], *Pseudomonas aeruginosa* [16], *Pseudomonas fluorescens* [17], *Candida rugosa* [18], *Candida antarctica* [19], *Thermomyces lanuginosus* [20] and *Burkholderia cepacia* [21].

Lipases are produced in industrial scale mostly by submerged fermentation, where the medium consists of a liquid containing various solubilized nutrients. Lipases can also be produced by solid state fermentation (SSF). In this process a solid substrate is used as support and the growth and metabolism of microorganisms occur in the absence or near-absence of free water [22]. The application of solid substrates is economically important for countries with an abundance of biomass and agroindustrial residues, which can be used as cheap raw materials [23].

Since lipases have several applications and have been the focus of numerous research studies, there is a need for standardized methods in order to compare the catalytic efficiency of lipases obtained from different sources in distinct laboratories. There are several methods available for measuring lipolytic activity as well as for the detection of lipases. These methods can be classified as follows: 1) titrimetry; 2) spectroscopy (photometry, fluorimetry, infra-red); 3) chromatography; 4) radioactivity; 5) interfacial tensiometry; 6) turbidimetry; 7) conductimetry; 8) immunochemistry and 9) microscopy [24]. In this chapter, however, we describe only the main qualitative, semi-quantitative and quantitative lipase activity assays using soluble and insoluble substrates. Due to the fact that some methods can be used either for lipase or for esterases, some methods will be described for both enzymes.

Screening for Lypolitic Microorganisms

Qualitative assays are important tools that can be used in preliminary studies aiming the identification of microorganisms that produce lipases and esterases. They are useful for screening large numbers of microorganisms for enzyme activity, when quantitative methods are not required, since these last are expensive and time consuming.

Screening for microorganisms that produce lipase and esterase can be accomplished through the use of a selective agar medium containing lipids as carbon source. After inoculating and incubating the microorganism on the medium, the enzyme is produced and it hydrolyzes lipids, forming hydrolysis zones around microbial growth. According to the literature, the measurement of hydrolysis zones can be used as a semi-quantitative determination of the presence of the enzyme. These zones can have different aspects depending on the methodology used [25]. There is a range of lipid sources and methodologies that can be applied. The screening is very important for posterior study and application of lipases. Then, most of groups that study lipases follow the steps described below:

1. Screening of microorganisms that produce lipases by the determination of halos developed around microorganism colonies grown on medium containing lipid sources (qualitative assays).

2. Estimation of the enzyme activities semi-quantitatively through calculation of the enzymatic index R/r, where R is the radius of the clear or opaque or colored areas (in cm), developed during incubation time, and r is the colony radius (in cm).

3. In the last step, the best producers selected in the previous steps are inoculated in an appropriated fermentation medium for lipases production and at the end of the bioprocess, enzyme activities are measured using quantitative methods.

Lipases are activated when adsorbed to an oil-water interface and do not hydrolyze dissolved substrates efficiently, displaying little activity in aqueous solutions containing soluble substrates, in contrast to esterases [1]. Then, a 'true' lipase hydrolyzes emulsified esters of glycerol and long-chain fatty acids such as triolein, trioctanoin, tripalmitin and olive oil, although they also display activity against emulsions water-soluble short-chain TAG (triacylglycerol) such as tripropionin and tributyrin. On the other hand, esterases are defined as enzymes that act on solutions of short-chain fatty acyl esters such as methyl butyrate, ethyl butyrate, triacetin and tributyrin, whereas possess no activity against emulsions of long-chain TAG [7, 26]. The surfactant Tween has been used both to screen lipases and esterases [9, 24].

Due to the growing interest in synthetic biocatalytic transformations catalyzed by lipases, this topic will describe not only the selection of microorganisms with hydrolysis activity, but also strains with synthesis activity.

There are some materials and equipment which are used in all screening methods and are listed below; the ones specific to each method will be listed later. For each method, examples of medium composition and incubation conditions will be given.

Materials

1.   Nitrogen sources;

2.   Agar-agar;

3.   Sodium hydroxide or hydrochloric acid for pH adjust;

4.   Enzyme inducer (triacylglycerols - TAG);

5.   Beakers;

6.   Petri dishes;

7.   Platinum inoculating loop.

Equipment

1.   Analytical balance;

2.   pH meter;

3.   Autoclave;

4.   Incubation chamber;

5.   Homogenizer;

6.   Chamber with UV radiation.

Screening Methods for Hydrolysis Activity of Lipases

Usually, the methodologies used to determine lipase/esterase activities are similar. Unfortunately, several are unspecific for lipases or esterases, since some substrates can be hydrolyzed by both enzymes.

Detection of Lipolytic Microorganisms by the Formation of Opacity Zones

This method is based on the capacity of microorganisms to produce lipases and esterases. The hydrolysis of the TAG generates an opacity zone around microbial growth on agar plate, due to the formation of calcium salts.

Materials

1. Strains of interest;

2. Agar medium containing TAG and $CaCl_2$.

    e.g., (% w v^{-1} in distilled water): glucose (2.0), yeast extract (0.5), peptone (0.5), agar-agar (2.0), supplemented with Tween 80 (1.0) and $CaCl_2$ (0.1). The media pH is adjusted to 4.5 for yeasts and to 6.0 for fungal strains [9].

Procedure

1. Prepare agar medium plates containing TAG. After autoclaving, cool it to about 60 °C and emulsify with a homogenizer sterilized by UV radiation, mixing for at least 2 min. Distribute the agar into the petri dishes.

2. Inoculate microorganisms as spots on the agar surface and incubate them in adequate conditions. This test should preferably be carried out using triplicates.

    e.g., Aspergillus oryzae, A. niger and *Rhizopus nigricans* were inoculated 5 days at 30 °C; *Candida rugosa, Candida lipolytica* and *Hansenula polymorpha* were incubated 3 days at 30 °C [9].

3. Detect microorganisms that produce lipase/esterase by the appearance of an opacity zone surrounding microbial growth. These opacity zones occur due to the formation of calcium salts. According to Ionita *et al.* [9], this technique can be used to classify the microorganisms as weak (R/r < 1.0), moderate (1.0 < R/r < 2.0) or good enzyme producers (R/r > 2.0).

Detection of Lipolytic Microorganisms by the Formation of Clear Zones

This method is based on the capacity of microorganisms to produce lipases and esterases and consists basically of determining the diameter of the product diffusion area. The hydrolysis of TAG added previously in the agar plate will give rise to a clarified zone around microbial growth (Fig. **2**). This detection method is proposed mainly for microorganisms that produce esterase, due to the optical phenomenon that is observable only if the released fatty acids are partly water-soluble, which are derived from the hydrolysis of short-chain TAG [24]. However, this method has already been described in the literature to select microorganisms that produce lipases using Tween [24] and tributyrin [27].

Materials

1. Strains of interest;

2. Agar medium containing the TAG;

   e.g., (% w v^{-1} in distilled water): peptone (0.5), yeast extract (0.3), agar-agar (2.0), supplemented with tributyrin (0.1). The medium pH was adjusted to pH 6.0 [28].

Procedure

1. Prepare agar medium plates containing TAG. After autoclaving, cool it to about 60 °C and emulsify with a homogenizer sterilized by UV radiation, mixing for at least 2 min. Distribute the agar into the petri dishes.

2. Inoculate microorganisms as spots on the agar surface and incubate them in adequate conditions. This test should preferably be carried out using triplicates.

   e.g., fungal strains were incubated 7 days at 30 °C [27].

3. Detect microorganisms that produce lipase/esterase by the formation of a clear zone surrounding microbial growth. The performance of the microorganisms can be determined evaluating the value of the enzymatic index *(R/r)* formed after the incubation.

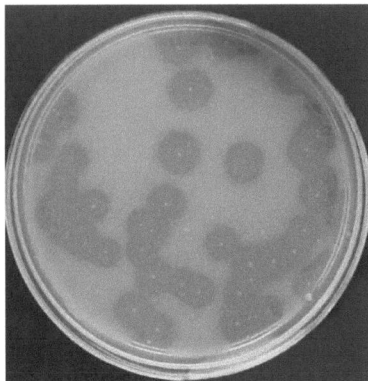

Figure 2: Agar plate with growth of microbial strains in a medium containing tributyrin, showing the hydrolysis zones generated by lipase/esterase activity.

Detection of Lipolytic Microorganisms by the Formation of Color Zones Using Dyes

This method is based on the capacity of microorganisms to produce lipases and esterases. The hydrolysis of the TAG is monitored through the change in the color of pH indicators due to the fatty acids released by the enzymatic action [24]. The examples of dyes that can be added are Victoria blue B, Spirit blue, Nile blue sulphate, night blue and others. Although this technique is very practical for screening of lipolytic microorganisms, the production of other acid metabolites than free fatty acids can give false results [15].

Materials

1. Strains of interest;

2. Agar medium containing TAG and pH indicators;

   e.g., (% w v^{-1} in distilled water): nutrient broth (0.8), agar-agar (1.0), supplemented with triolein (0.25) and Victoria Blue (0.01). The media pH was adjusted to pH 7.0 for the bacteria strains [29].

Procedure

1. Prepare agar medium plates containing the TAG and the dye. After autoclaving, cool it to about 60 °C and emulsify the medium mixing for at least 2 min with a homogenizer sterilized by UV radiation.

2. Inoculate microorganisms as spots on the agar surface and incubate in adequate conditions. This test should preferably be carried out using triplicates.

   e.g., Geobacillus zalihae was incubated at 60 °C [29].

3. Detect microorganisms that produce lipase/esterase by observing a change in color of the dye added.

Detection of Lipolytic Microorganisms Using Rhodamine B

Screening of lipolytic microorganisms on agar plates is frequently performed using tributyrin as a substrate. However, this substrate is not suitable to detect true lipases because it is also hydrolyzed by esterases [30]. Alternatively, Rhodamine B plate method using as substrates long-chain TAGs is a specific and sensitive assay to identify microorganisms that produce lipases. Therefore, a strategy frequently used is, at first, screening the microorganisms using tributyrin as a substrate aiming to verify the lipolytic potential producers and then to apply Rhodamine test with the positive ones to verify the presence of true lipases.

The addition of Rhodamine B allows the agar plate containing TAG to become pink, and then the positive colonies can be identified by the orange fluorescent halo formed around them. Such halos can be clearly seen under UV light at 350 nm [28]. On the other hand, the negative strains do not form orange fluorescence halo upon UV irradiation. The mechanism of the fluorescent products formation generated from TAG hydrolysis by lipases in Rhodamine B agar plate was not elucidated. A feasible mechanism may be the generation of excited dimers of Rhodamine, which emit fluorescence. Other characteristics of the Rhodamine B plate method are its insensibility to pH changes and to exopolysaccharide production. Besides this, as no growth inhibition or any change in microorganism physiological properties occurs, it allows its re-isolation, giving to the method a greater aplicability [30].

Some researchers have opted to use this method for screening true lipase producers, nevertheless until now there is no consensus about the dye concentration that must be added to the agar plate. The most common values range from 0.001 % w v^{-1} to 2.0 % w v^{-1} [28; 30-33].

Materials and Equipment

1. Chamber with UV radiation;

2. Strains of interest;

3. Agar medium containing long-chain TAG and Rhodamine B.

    e.g., (% w v^{-1} in distilled water): agar-agar (1.5), supplemented with olive oil (2.0) [33]. The solution of Rhodamine B in distilled water is prepared separately [30] and the concentration should be sufficient to ensure the final concentration of 0.007 % w v^{-1} in the agar plate [33].

Procedure

1. Prepare agar medium plates containing TAG. After autoclaving, cool it to about 60 °C mixing with the Rhodamine B solution sterilized by microfiltration. The mixture must be then emulsified by mixing for at least 2 min with a homogenizer sterilized by UV radiation.

2. Inoculate microorganisms as spots on the agar surface and incubate them in adequate conditions. This test should preferably be carried out using triplicates.

    e.g., fungal strains were incubated 2 days at 30 ºC [33].

3. Detect microorganisms that produce lipase by observing the formation of orange fluorescent halos around the positive colonies, once the plates are irradiated with UV light at 350 nm (Fig. **3**). Classify lipase producers as follows: the enzymatic index (*R/r*) of 1 means no hydrolysis, thus no lipase production, while a higher (*R/r*) ratio suggests lipase production [34].

Comments

Some screening studies have been carried out aiming the identification of stereospecific lipases and esterases. The methods described above can also be used for this kind of screening, replacing the substrate by a specific one [35].

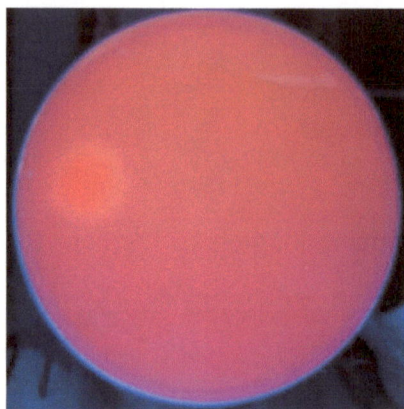

Figure 3: Agar plate with growth of microbial strain in a medium containing olive oil and Rhodamine B, showing the fluorescence zone generated by lipase activity. Two different microbial strains were incubated in each side of the plate, but only the left side strain was positive. No growth was detected for the right side strain.

Some works describe the correlation between halos zones (opaque, clear or colored) formed by the methods quoted above and the lipase activity. A direct quantitation of lipase activity is difficult, due to the low amounts of lipase molecules released by a colony. However, the quantification can be done, by filling enzyme culture supernatants with known activity into holes punched into the agar containing TAG. The samples for filling must have the same volume but different activities. The zone diameter (diameter of the halo minus the diameter of the hole) is linearly related to the logarithm of lipase activity. Then, this approach enables the construction of an equation that allows the estimation of the correlation between enzyme activity and halos diameters [24, 25, 30]. Each determination should be an average from triplicate samplings. According to Kouker and Jaeger [30], the sensitivity of the assay is demonstrated by the fact that the lowest detection limit is about twenty times lower than the required by the titrimetric assay.

Screening Method for Synthesis Activity of Lipases Using Rhodamina B

This method is specific for lipases and is based on the esterification of a long-chain alcohol and a fatty acid in Rhodamine B agar plates. In the presence of a lipase an ester is formed and the fluorescence of the rhodamine–fatty acid complex under UV light vanishes. To perform such analysis, limits to water activity and lipase activity must be observed. The upper bound limit to water activity has been shown to be 0.7, while the lower bound limit of lipase activity was demonstrated to be 60 mU. Values within such ranges can be used with no further restrictions. The ester synthesis can be confirmed using HPLC [36].

Materials and Equipment

1. Chamber with UV radiation;

2. Strains of interest;

3. Agar medium containing the alcohol and the fatty acid;

   e.g., in distilled water: nutrient broth (0.80 % w v^{-1}), sodium chloride (0.40 % w v^{-1}) and agar-agar (1.50 % w v^{-1}), supplemented with oleic acid (1.56 % v v^{-1}) and dodecanol (1.56 % v v^{-1}) [36]. The solution of Rhodamine B is prepared separately and the concentration should be sufficient to ensure the final concentration of 0.001 % w v^{-1} in the agar plate [30].

Procedure

1. Prepare agar medium plates containing the fatty acid and alcohol. After autoclaving, cool it to about 60 °C mixing with the Rhodamine B solution sterilized by microfiltration. The mixture must be emulsified mixing for at least 2 min with a homogenizer sterilized by UV radiation.

2. Inoculate microorganisms as spots on the agar surface and incubate in adequate conditions. This test should preferably be carried out using triplicates.

   e.g., Escherichia coli mutant carrying the *Rhizopus delemar* lipase gene was incubated 48 h at 37 °C [36].

3. Detect microorganisms that produce lipases by observing the formation of a dark halo under UV light.

4. The presence of the reaction product (ester) can be confirmed by cutting and dissolving the dark halos with a Gel Extraction kit. After

mixing the extracted sample with hexane, an organic phase is obtained. The hexane must be evaporated and the products analyzed by HPLC [36].

Comments

Besides the Rhodamine method, Konarzycka-Bessler and Bornscheuer [37] and Schmidt and Bornscheuer [38] describe a method based on the transesterification between an alcohol and a vinyl ester of a carboxylic acid in the presence of an organic solvent. However, this method has some problems, which difficult its use: water must be removed, derivatization of substrates is required and it is not specific for lipases [36].

Assays for Lipase Activity

After the screening of microorganisms that produce lipases, the best producers can be inoculated in a fermentation medium for lipases production and the enzymatic activities can be measured using quantitative methods.

Among the several methods described in the literature to measure hydrolysis or synthesis activity of lipases [24, 25, 38], the titrimetric and spectrophotometric methods are the most common used, therefore, in this chapter we only describe these.

Measurement of Hydrolysis Activity of Lipases

In this section three most common assays for lypolytic activity will be reported. Two of them are based on the measurement of fatty acids released in the triacylglycerol hydrolysis by lipases, whether by titration of the liberated acid (titrimetric method) or detection of the fatty acids by complexation with a colorimetric reagent of cupric acetate (colorimetric copper soap method). The other one is based on the *p*-nitrophenyl acyl esters hydrolysis and quantify the chromogenic alcohol released, providing for a continuous spectrophotometric assay (spectrophotometric *p*-nitrofenil palmitate method).

Determination of Hydrolysis Activity of Lipases by the Titrimetric Method

The titrimetric method to determine the lipase hydrolytic activity is based on the use of different triacylglycerols which are hydrolyzed into fatty acids during enzymatic reaction. These released acids are titrated with sodium hydroxide to an end point.

Materials

1. Gum Arabic;

2. Long-chain TAG emulsion (substrate);

3. 200 mM sodium phosphate buffer, pH 7.0;

4. 0.02 N sodium hydroxide solution;

5. Ethanol: acetone: water solution;

6. Potassium hydrogen phthalate;

7. Flask of glass or plastic;

8. Lipase.

Equipment

1. Analytical balance;

2. Magnetic stirrer;

3. Homogenizer;

4. Water bath with temperature controller;

5. Automated pH stat titrator.

Procedure

Reagents and Solutions

Long-Chain TAG Emulsion [39]

At room temperature, weight 7 g of gum arabic in a beaker. Add to 7 g of gum arabic 50 mL of long-chain TAG (*e.g.,* extra virgin olive oil) and 50 mL of distilled water. Mix for approximately 3 min, until a homogenous suspension is obtained (Fig. **4A**).

Obs.: The suspension should be prepared immediately before starting the enzymatic reaction.

Potassium Hydrogen Phthalate Solution

At room temperature, weight 0.2 g of potassium hydrogen phthalate previously dried into a plastic or glass flask. Add 15 mL of distilled water and solubilize the solid using a magnetic stirrer.

Preparation and Standardization of Sodium Hydroxide Solution

At room temperature, weight 0.8 g of sodium hydroxide. Add 500 mL of distilled water and solubilize the hydroxide. Bring the volume to 1000 mL with distilled water in a volumetric flask. Store the solution in plastic bottle.

Standardize the sodium hydroxide solution with potassium hydrogen phthalate solution using the Automated pH stat titrator.

Ethanol: Acetona: Water solution (1:1:1)

Add the same volume of ethanol, acetone and water to a flask. Homogenize. Prepare this solution a few minutes before starting the reaction.

Assays

1. Place 4 mL of buffer and 5 mL of emulsion into a flask. Preincubate 1 min in agitated water bath at specific temperature depending on each enzyme.

   e.g., Damaso *et al.* [15] used 35 °C for a mesophilic enzyme.

2. Add 1 mL of the lipase to the flask and homogenize. If the enzyme is in solid state, add 1 g. If necessary, the liquid enzymes can be diluted and considering the solid ones, the amount can be reduced.

3. Start timer and keep stirring the mixture in the same temperature of the preincubation.

4. Stop the reaction using 10 mL of the ethanol:acetone:water solution after an incubation time that ensure that the reaction is still in the initial velocity.

    Obs.: This test should preferably be carried out in triplicates.

5. Titrate the contents of each flask with 0.02 N sodium hydroxide solution using an automated pH stat titrator until pH 11.0 [40]. The titrator electrode should be specific for non-aqueous media. If it is necessary, add 10 mL or more of distilled water in each flask to reach the volume that ensures that the electrode is submerged in the mixture (Fig. **4**B).

6. Blank assays should be conducted the same way, but adding the enzyme just before titration.

    Obs.: This blank test should preferably be carried out using duplicates.

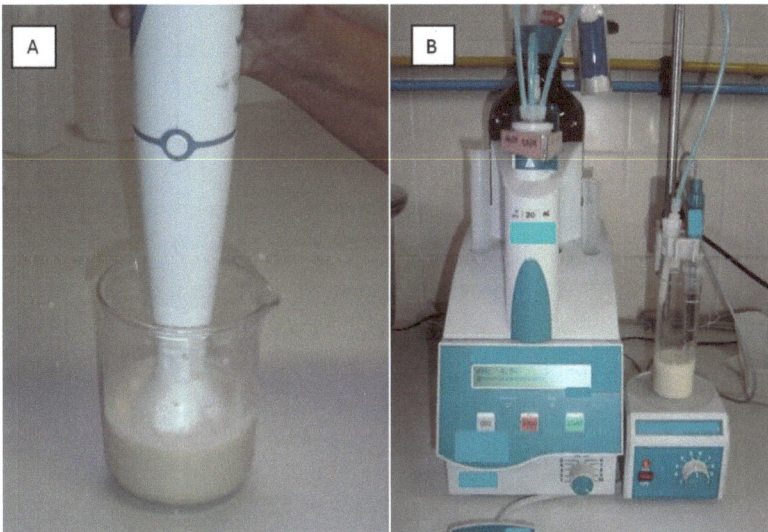

Figure 4: A. Preparation of homogenous TAG emulsion using the mixer; **B.** Titration of enzymatic reaction products in the pH-stat titrator.

Analyze Data

Determine the number of units (U) of enzymatic activity, which is defined as the amount of enzyme that produces 1 μmol of fatty acids per minute under the specified assay conditions. The calculation of the quantity of fatty acids released in the sample is based on the difference of equivalents of sodium hydroxide used to reach the end point of the enzyme reaction sample and the blank one, as follows below (Equation 1):

$$Activity \left(\frac{U}{mL} \right) = \frac{\lfloor (V_s - V_b) \times N \times 1000 \rfloor}{(t \times V_{enz})} \tag{1}$$

> where V_s is the sodium hydroxide volume used to titrate the sample; V_b is the sodium hydroxide volume used to titrate the blank; N is the normality of the sodium hydroxide titrant used (0.02 N in this case); t is the time reaction; V_{enz} is the volume of enzyme used in the reaction. Multiplication by 1000 is to adjust the units.

The use of the buffer that was described above for the enzyme reaction and the titrant normality of 0.02 N for sodium hydroxide in this test are suggested when the lipase is considered neutral. If the enzyme has an acidic or basic characteristic, other conditions need to be used. *E.g.,* Sodium citrate buffer 50 mM, pH 4.0 and sodium hydroxide solution 0.05 N for acidic lipases [32].

It is interesting to note that if the enzyme is in solid state, the V_{enz} equation term will be replaced to M_{enz} (mass of enzyme in grams) and the activity will be defined in terms of U g^{-1}. Furthermore, it is possible to calculate the specific activity when the enzyme was purified or to calculate apparent specific activity if it was a crude extract. Both kinds of specific activities (U mg^{-1} protein) can be calculated dividing the lipase activity (U mL^{-1}) by the amount of protein (in mg mL^{-1}) present in enzymatic extract. The protein content can be measured using methods such as Lowry [41] or Bradford [42].

Comments

The procedure presented here for measurement of lipase activity was adapted from Pereira *et al.* [39]. The modifications made in the original methodology are concerning the method to measure the amount of sodium hydroxide spent to reach the end point and the pH value of this point, which was replaced from 7.0 to 11.0 [40]. The use of pH stat titrator replaces the manual and visual method, which uses the burette and the pH indicator as phenolphthalein. This modification in the procedure is quite advantageous because the manual/visual method (burette) is not precise nor accurate, labor intensive and time consuming, although it has been frequently used with phenolphthalein or thymolphthalein indicators [9, 43, 44].

An important point to be stressed is that sometimes it is advisable to dilute the enzyme before analysis. This reduce the time of titration analysis, the use of sodium hydroxide (lowering costs) and the amount of waste to be treated.

Other methods for the titration step are described in literature. Pinsirodom and Parkin [44] remove samples of reaction mixture at five suitable intervals (*e.g.,* 5, 10, 15, 20, and 25 min), and transfer each sample to a separate flask for carry out the titulation through the manual method using phenolphthalein. This strategy enables the construction of a fatty acid released curve and the calculation of the hydrolitic activity at the initial velocity range. Furthermore, Salum *et al.* [21] placed a reaction mixture in a thermostated vessel and the release of free fatty acids was monitored continuously during the reaction (5 min) by titration, using a pH-Stat titrator set at pH 7.0. This titration method still allows calculating the lipase activity at the initial velocity, as described above.

Even though esterases and lipases differ in their action ability, some assays presented here can be used to measure both types of enzyme activities, since they share the general ability to hydrolyze carboxyl esters of various alcohols.Water-soluble substrates (short-chain TAG) can be used to quantified esterase activity [7, 26] and the use of water-insoluble substrates (long-chain TAG) is considered diagnostic for lipases as presented in this procedure.

Determination of Hydrolysis Activity of Lipases Using the Colorimetric Copper Soap Method

In this method [44] fatty acids released during hydrolysis of olive oil or triolein can be quantified colorimetrically using a cupric acetate/pyridine reagent. Fatty acids complex with copper to form copper soaps, which absorb light in 715 nm, yielding a blue color. Quantification of fatty acids is determined by reference to a standard curve prepared using oleic acid. This colorimetric determination of fatty acids was developed by Lowry and Tinsley [45].

Materials

1. 25 mM oleic acid standard solution;

2. Benzene;

3. Cupric acetate-pyridine reagent;

4. Olive oil/ Triton X-100 emulsion (substrate);

5. 50 mM sodium phosphate buffer, pH 8.0;

6. Glass cuvettes;

7. Lipase.

Equipment

1. Analytical balance;

2. Spectrophotometer;

3. pH meter;

4. Homogenizer;

5. Vortex mixer;

6. Magnetic stirrer;

7. Water bath;

8. Centrifuge.

Procedure

Reagents and Solutions

Cupric Acetate-Pyridine Reagent

Prepare a 5 % (w v^{-1}) aqueous solution of cupric acetate, filter and adjust the pH of the solution to 6.0 to 6.2 using pyridine. Store at room temperature.

Olive oil/Triton X-100 Emulsion

Dissolve 5 g each olive oil and Triton X-100 in 5 mL of chloroform. Evaporate the chloroform under a stream of nitrogen gas at 60 °C until the mixture becomes turbid (approximately 15 min). Slowly add 50 mM sodium phosphate buffer, pH 8.0 at 60 °C while swirling the flask and bring to 100 mL. Adjust the pH to 8.0 using NaOH. Prepare fresh daily and rehomogenize the mixture periodically during the entire day of use.

Standard Curve

Place 0.1 to 1.0 mL of 25 mM oleic acid standard solution in tubes and dilute each to 5 mL with benzene for the standard curve (final 2.5 to 25 μmol oleic acid). These solutions should preferably be carried out in triplicates;

Add 1 mL cupric acetate/pyridine reagent to each tube and vortex 2 min. Centrifuge 5 min at 1000 × g;

Measure the A$_{715}$ of the organic layer against a benzene blank and construct a standard curve by plotting A$_{715}$ *vs.* the amount of oleic acid in 5 ml benzene.

Assays

To prepare the reagent blank add 0.3 mL olive oil/Triton X-100 emulsion to a tube containing 5 mL of benzene and 1 mL cupric acetate/pyridine reagent. Vortex 2 min and centrifuge 5 min at 1000 × g;

Place 25 mL olive oil/Triton X-100 emulsion into an Erlenmeyer flask. Preincubate 15 min with magnetic stirring in a water bath at 37 °C (or the appropriate temperature for the lipase used);

Add the lipase (limit to 0.5 mL in solution form) to initiate hydrolysis. Remove 0.3 mL samples of the reaction mixture at predetermined time intervals and place in tubes containing 5 mL of benzene and 1 mL cupric acetate/pyridine reagent. Vortex 2 min and centrifuge 5 min at 1000 × g;

Zero the spectrophotometer at 715 nm using the organic layer of the blank. Measure the A_{715} for the organic layer of each sample.

Obs.: This test should preferably be carried out in triplicates.

Analyze Data

1. Convert A_{715} values to µmol oleic acid/mL sample using the equation of the line of the standard curve as follows (Equation 2):

$$\frac{\mu mol\ oleic\ acid}{\mu L\ sample} = \frac{\left(A_{715} - y\ intercept\right)}{slope \times 0.3\ mL\ sample} \tag{2}$$

2. Compose a reaction progress curve by plotting the concentration of oleic acid *vs.* reaction time. Draw a tangent to the initial (and linear) portion of the progress curve to obtain initial reaction rates (v_0) as follows (Equation 3):

$$v_0 = slope = \frac{\left(y_2 - y_1\right)}{\left(x_2 - x_1\right)} \tag{3}$$

3. Determine specific activity of the lipase preparation as follows (Equation 4a):

$$Specific\ activity \left(\frac{U}{mg\ protein}\right) = \frac{v_0}{\left(\dfrac{a}{25+b}\right)} \tag{4a}$$

where a is the amount of protein added (in mg) to the reaction mixture; b is the volume of the enzyme added (mL) to the reaction mixture, and $25 + b$ is the total volume of the reaction mixture (in mL).

One unit of activity (U) is defined as the amount of enzyme that produces 1μmol of fatty acids per minute under the conditions of the assay. Activity can be expressed in U (μmol min^{-1}) per mg protein, or as U mL^{-1} if the enzyme is in a liquid form and the concentration of protein is unknown. In the last case, use the Equation 4b:

$$Activity \left(\frac{U}{mL} \right) = \frac{v_0}{\left(\dfrac{b}{25+b} \right)} \tag{4b}$$

Comments

Some authors use other solvents instead of benzene, such as toluene [46] and isooctane [47, 48].

Determination of Hydrolysis Activity of Lipases Using the Spectrophotometric p-Nitrofenil Palmitate (pNPP) Method

This method quantifies the level of *p*-nitrophenol (A$_{max}$ 410 nm, yellow color) released following the hydrolysis of *p*-nitrophenyl palmitate (*p*NPP) substrate by lipase. The method reported below was described by Winkler and Stukmann [49] and has been modified by other authors.

Materials

1. *p*-nitrophenol (*p*NP) standard solutions;

2. Solution A: *p*-nitrophenyl palmitate (*p*NPP) in isopropanol solution;

3. Solution B: Triton X-100 and gum arabic in 50 mM phosphate buffer, pH 8.0;

4. 50 mM sodium phosphate buffer, pH 8.0;

5. Cuvettes;

6. Lipase.

Equipment

1. Analytical balance;

2. Spectrophotometer;

3. pH meter;

4. Magnetic stirrer;

5. Water bath.

Procedure

Reagents and Solutions

Solution A

Prepare a 3 mg mL^{-1} pNPP solution in isopropanol. Heat the flask (37 °C) to facilitate the dissolution. Wrap the flask in aluminum foil to protect from light and store in freezer.

Solution B

Prepare a solution containing 4.44 g L^{-1} Triton X-100 and 1.11 g L^{-1} gum arabic in 50 mM phosphate buffer, pH 8.0. Store at 4 °C.

Standard Curve

Prepare solutions with concentrations of 7-70 μmol L^{-1} of pNP in Solution B. These solutions should preferably be carried out in triplicates.

Measure the A$_{410}$ of the solutions against a Solution B blank and compose a standard curve by plotting A$_{410}$ vs. the pNP concentration (in mol L^{-1}).

Assays

Slowly, add 1 part of Solution A to 9 parts of Solution B into a beaker, under magnetic stirring. This solution will be called Solution C. The mixture should be made immediately before use, since the substrate is unstable in aqueous solution;

Prepare the reagent blank by adding into a cuvette 0.9 mL of Solution C and 0.1 mL of 50 mM phosphate buffer, pH 8.0;

Zero the spectrophotometer at 410 nm using the reagent blank;

Into a cuvette, add 0.9 mL of Solution C and 0.1 mL of lipase;

Measure A_{410} for the sample at 37 °C (or the appropriate temperature for the lipase used) in spectrophotometer with controlled heating in cell compartment. Follow the reaction kinetics measuring A_{410} at predetermined time intervals (*e.g.,* 0, 10, 20, 30, 40, 50 and 60 s).

Obs.: This test should preferably be carried out in triplicates.

Analyze Data

Use the standard curve to calculate the molar absorption coefficient as follows (Equation 5):

$$A = \varepsilon \times b \times c \tag{5}$$

> where A is absorbance, ε is the molar absorption coefficient, b is the path length and c is the concentration of *p*-nitrophenol. Since we use path length = 1, ε is the angular coefficient of the equation of the line of the standard curve. The ε unit is L mol^{-1} cm^{-1}.

With the data obtained in the assay, construct a reaction progress curve by plotting the absorbance *vs.* reaction time (in min). Draw a tangent to the initial (and linear) portion of the progress curve to obtain the angular coefficient (*ac*) as follows (Equation 6):

$$ac = slope = \frac{\left(y_2 - y_1\right)}{\left(x_2 - x_1\right)} \qquad (6)$$

Calculate the lipolytic activity as follows (Equation 7):

$$Activity \left(\frac{U}{mg\ protein}\right) = \frac{\left(ac \times d \times 10^6\right)}{\left(\varepsilon \times 10^3 \times 10^{-1} \times e\right)} \qquad (7)$$

where *ac* is the angular coefficient, *d* is the dilution factor of the enzyme, ε is the molar absorption coefficient and *e* is the concentration of lipase added (mg protein mL^{-1}). Multiplication by 10^6 is to obtain the value in µmoL, 10^3 is to give the activity in 1 mL (reactional volume) and 10^{-1} is the volume of the lipase added (100 µL). Then, it is possible to obtain the Equation 8a:

$$Activity \left(\frac{U}{mg\ protein}\right) = \frac{\left(ac \times d \times 10^4\right)}{\left(\varepsilon \times e\right)} \qquad (8a)$$

One unit of activity (U) is defined as the amount of enzyme that produces 1µmol of p-nitrophenol per minute under the conditions of the assay. Activity can be expressed in U (µmol min^{-1}) per mg protein, or as U mL^{-1} if the enzyme is in a liquid form and the concentration of protein is unknown. In the last case, use the Equation 8 b.

$$Activity \left(\frac{U}{mL}\right) = \frac{\left(ac \times d \times 10^4\right)}{\varepsilon} \qquad (8b)$$

Comments

Other substrates can be used besides *p*-nitrophenyl palmitate, such as *p*-nitrophenyl laurate, *p*-nitrophenyl acetate and *p*-nitrophenyl butyrate. Use of a shorter chain length *p*-nitrophenyl acyl ester, especially the water-soluble acetate derivative, would provide a preliminary indication of esterase-type activity of an unknown preparation.

Measurement of Synthesis Activity of Lipase Using the Colorimetric Copper Soap Method

The application of lipases in non-aqueous media in industrial sectors such as food and pharmaceutical, becomes the analytical methods to determine the synthesis activity of these enzymes so important nowadays.

The colorimetric assay using the copper soap method can also be applied for esterification reactions. In this case, lipases should be lyophilized or immobilized, since the water in the medium favor the reverse reaction (hydrolysis). The next method that will be described is the determination of ethyl oleate synthesis activity [50].

Fatty acids consumed during ethyl oleate synthesis reaction can be quantified colorimetrically using a cupric acetate/pyridine reagent. Fatty acids complex with copper to form copper soaps that absorb light in 715 nm, yielding a blue color. Quantification of fatty acid consumed is determined by reference to a standard curve prepared using oleic acid. This colorimetric determination of fatty acids was developed by Lowry and Tinsley [45].

Materials

1. 25 mM oleic acid standard solution;

2. Toluene;

3. Cupric acetate-pyridine reagent;

4. Oleic acid;

5. Ethanol;

6. *n*-heptane;

7. Glass cuvettes;

8. Lipase.

Equipment

1. Analytical balance;

2. Spectrophotometer;

3. pH meter;

4. Magnetic stirrer;

5. Water bath;

6. Centrifuge.

Procedure

Reagents and Solutions

Cupric Acetate-Pyridine Reagent

Prepare a 5 % (w v^{-1}) aqueous solution of cupric acetate, filter and adjust pH of the solution to 6.0 to 6.2 using pyridine. Store at room temperature.

Standard Curve

Place 0.1 to 1.0 mL of 25 mM oleic acid standard solution in tubes and dilute each to 5 mL with toluene for the standard curve (final 2.5 to 25 µmoL oleic acid). These solutions should preferably be carried out in triplicates;

Add 1 mL cupric acetate/pyridine reagent to each tube and vortex 2 min. Centrifuge 5 min at 1000 × g;

Measure the A$_{715}$ of the organic layer against a toluene blank and construct a standard curve by plotting A$_{715}$ vs. the amount of oleic acid in 5 mL toluene.

Assays

Into an Erlenmeyer flask, place 210 mM ethanol in 25 mL *n*-heptane. Preincubate 15 min with magnetic stirring in a water bath at 37 °C (or the appropriate temperature for the lipase used). Add the lipase, homogenize, remove a 0.3 mL sample (reagent blank) and place in a tube containing 5 mL of toluene and 1 mL cupric acetate/pyridine reagent. Vortex 2 min and centrifuge 5 min at 1000 × g;

Add oleic acid (70 mM) to the Erlenmeyer flask to initiate the esterification. Remove 0.3 mL samples of the reaction mixture at predetermined time intervals (initiating by 0 min) and place in tubes containing 5 mL of toluene and 1 mL cupric acetate/pyridine reagent. Vortex 2 min, centrifuge 5 min at 1000 × g;

Zero the spectrophotometer at 715 nm using the organic layer of the blank. Measure the A_{715} for the organic layer of each sample.

Obs.: This test should preferably be carried out in triplicates.

Analyze Data

Convert A_{715} values to μmol oleic acid/mL sample using the equation of the line of the stardard curve as follows (Equation 9):

$$\frac{\mu mol\ oleic\ acid}{\mu L\ sample} = \frac{(A_{715} - y\ intercept)}{slope \times 0.3\ mL\ sample} \tag{9}$$

Compose a reaction progress curve by plotting the concentration of oleic acid *vs.* reaction time. Draw a tangent to the initial (and linear) portion of the progress curve to obtain initial reaction rates (v_0) as follows (Equation 10):

$$v_0 = slope = \frac{(y_2 - y_1)}{(x_2 - x_1)} \tag{10}$$

Determine specific activity of the lipase preparation as follows (Equation 11a):

$$Specific\ activity\ \left(\frac{U}{mg\ protein}\right) = \frac{v_0}{\left(\dfrac{a}{25}\right)} \tag{11a}$$

where a is the amount of protein added (in mg) to the reaction mixture and 25 is the volume of the reaction mixture (in mL).

One unit of activity (U) is defined as the amount of enzyme that catalyze the consumption of 1 μmol of oleic acid per minute under the conditions of the assay. Activity can be expressed in U (μmol min^{-1}) per mg protein, or as U per mg support/immobilized lipase, if the enzyme is immobilized. In the last case, use the Equation 11b.

$$Activity\left(\frac{U}{mg\ support}\right)=\frac{v_0}{\left(\dfrac{b}{25}\right)} \tag{11b}$$

where b is the amount of support added (in mg) to the reaction mixture and 25 is the volume of the reaction mixture (in mL).

Comments

The synthesis activity of lipase could also be determined using titrimetric method. This method is based on the measurement of the fatty acid consumed in the esterification reaction by lipases as colorimetric one, although the fatty acid amount is quantified by titration with sodium or potassium hydroxide. Because water can not be present in the medium, the enzyme should be in solid form (lyophilized or immobilized). The esterification reaction is carried out in the thermostatized vessel with a fatty acid and an alcohol and the residual acid content is determined by titration [51].

Final Comments

The most common method of measurement of fatty acids released in triacylglycerols hydrolysis is the titrimetric method. Assays for lipase activity using the colorimetric copper soap method are similar to titrimetry, since both measure fatty acids released, however, the colorimetric method is more specific for fatty acids [44]. Recently, synthesis reactions using lipases have received attention due the several reactions that lipases have shown to be able to catalyze in organic medium. Therefore, some methodologies have been developed based in the determination of fatty acid consumption by titrimetry and colorimetry.

Among the spectrophotometric methods for lipase assay, the most used is based on using *p*-nitrophenyl acyl esters as substrate. Since lipases are classified as glycerol ester hydrolases (EC 3.1.1.3), by definition, their native substrate is a triacylglycerol. *P*-nitrophenyl palmitate (and other carboxylic acid esters) are model or 'synthetic' substrates. Other lipolytic enzymes, and some proteolytic enzymes, may hydrolyze *p*-nitrophenyl acyl esters but be inactive against triacylglycerols [44]. Then, a more laborious but reliable method for identifying a 'true' lipase is the determination of fatty acids released from a triacylglycerol, usually triolein [52]. However, after the enzyme is characterized as a lipase, *p*-nitrophenyl acyl esters are the indicated substrates to be used routinely to determine lypolytic activity, since this method is fast and can be carried out in small reaction volumes.

Equipment

Analytical balance;

pH meter;

Autoclave;

Incubation chamber;

Homogenizer;

Chamber with UV radiation;

Spectrophotometer;

Magnetic stirrer;

Vortex mixer;

Water bath with temperature controller;

Centrifuge;

Automated pH stat titrator.

ACKNOWLEDGEMENTS

The authors thank Embrapa, CAPES, FAPERJ, MCT CNPq for the financial support and the researchers Betania Ferraz Quirino and Thais Demarchi Mendes that kindly gave the photos.

CONFLICT OF INTEREST

The authors confirm that this chapter content has no conflict of interest.

REFERENCES

[1]   SHARMA, R.; CHISTI, Y.; BANERJEE, U. C. Production, purification, characterization, and applications of lipases. Biotechnol Adv 2001; 19(8): 627 - 662.

[2]   FERNANDES, M. L. M.; SAAD, E. B.; MEIRA, J. A.; RAMOS, L. P.; MITCHELL, D. A.; KRIEGER, N. Esterification and transesterification reactions catalysed by addition of fermented solids to organic reaction media. J Mol Catal B Enzym 2007; 44(1): 8–13.

[3]   BARON, A. M.; INEZ, M.; SARQUIS, M.; BAIGORI, M.; MITCHELL, D. A.; KRIEGER, N. A comparative study of the synthesis of n-butyl-oleate using a crude lipolytic extract of *Penicillum coryophilum* in water-restricted environments. J Mol Catal B: Enzym 2005; 34(1-6): 25-32.

[4]   JAEGER, K. E.; RANSAC, S.; DIJKSTRA, B. W.; COLSON, C.; HEUVEL, M. VAN; MISSET, O. Bacterial lipases. FEMS Microbiol Rev 1994; 15(1): 29-63.

[5]   OLLIS, D. L.; CHEAH, E.; CYGLER, M.; DIJKSTRA, B.; FROLOW, F.; FRANKEN, S. M.; *et al.* The α/β hydrolase fold. Protein Eng 1992; 5(3): 197-211.

[6]   VERGER, R. " Interfacial activation " of lipases: facts and artifacts. Trends Biotechnol 1997; 15(1): 32-38.

[7]   CHAHINIAN, H.; NINI, L.; BOITARD, E.; DUBÈS, J-PAUL; COMEAU, L-CLAUDE; SARDA, L. Distinction Between Esterases and Lipases: A Kinetic Study with Vinyl Esters and TAG. Lipids 2002; 37 (7): 653-662.

[8]   JAEGER, K-ERICH; EGGERT, T. Lipases for biotechnology. Curr Opin in Biotech. 2002; 13(4): 390-397.

[9]   IONITA, A.; MOSCOVICI, M; POPA, C.; VAMANU, A.; POPA, O.; DINU, L. Screening of yeast and fungal strains for lipolytic potential and determination of some biochemical properties of microbial lipases. J Mol Catal B Enzym 1997; 3(1-4): 147-151.

[10]   MAHADIK, N. D.; PUNTAMBEKAR, U. S.; BASTAWDE, K. B.; KHIRE, J. M.; GOKHALE, D. V. Production of acidic lipase by *Aspergillus niger* in solid state fermentation. Process Biochem 2002; 38(5): 715-721.

[11]   BURKERT, J.; MAUGERI, F.; RODRIGUES, M. I. Optimization of extracellular lipase production by *Geotrichum* sp. using factorial design. Bioresource Technol 2004; 91(1): 77-84.

[12]   VILLENEUVE, P. Plant lipases and their applications in oils and fats modification. Eur J Lipid Sci Tech 2003; 105(6): 308-317.

[13]   SHIMOKAWA, Y.; HIRATA, K-ICHI.; ISHIDA, T.; KOJIMA, Y.; INOUE, N.; QUERTERMOUS, T.; *et al.* Increased expression of endothelial lipase in rat models of hypertension. Cardiovasc res 2005; 66(3): 594-600.

[14]   HASAN, F.; SHAH, A. A.; HAMEED, A. Industrial applications of microbial lipases. Enzyme Microb Tech 2006; 39(2): 235–251.

[15]   DAMASO, M. C. T.; PASSIANOTO, M. A.; FREITAS, S. C. de; FREIRE, D. M. G.; LAGO, R. C. A.; COURI, S. Utilization of agroindustrial residues for lipase production by solid-state fermentation. Braz J Microbiol 2008; 39(4): 676-681.

[16]   RUCHI, G.; ANSHU, G.; KHARE, S. K. Lipase from solvent tolerant *Pseudomonas aeruginosa* strain: production optimization by response surface methodology and application. Bioresource Technol 2008; 99(11): 4796-4802.

[17]   KULKARNI, N.; GADRE, R. V. Production and properties of an alkaline, thermophilic lipase from *Pseudomonas fluorescens* NS2W. J Ind Microbiol Biot 2002; 28(6): 344 - 348.

[18]   MARIA, P. D. de; SANCHEZ-MONTERO, J. M.; ALCÁNTARA, A. R.; VALERO, F.; SINISTERRA, J. V. Rational strategy for the production of new crude lipases from *Candida rugosa*. Biotechnol Lett 2005; 27(7): 499-503.

[19]   LIU, J.; ZHANG, Y. Optimisation of lipase production by a mutant of *Candida antarctica* DSM-3855 using response surface methodology. Int J Food Sci Tech 2011; 46(4): 695-701.

[20]   JENSEN, B.; NEBELONG, P.; OLSEN, J.; REESLEV, M. Enzyme production in continuous cultivation by the thermophilic fungus *Thermomyces lanuginosus*. Biotechnol Lett. 2002; 24(1): 41-45.

[21]   SALUM, T. F. C.; VILLENEUVE, P.; BAREA, B.; YAMAMOTO, C. I.; CÔCCO, L. C.; MITCHELL, D. A. *et al.* Synthesis of biodiesel in column fixed-bed bioreactor using the fermented solid produced by *Burkholderia cepacia* LTEB11. Process Biochem 2010; 45(8): 1348-1354.

[22]   LONSANE, B. K.; GHILDYAL, N. P.; BUDIATMAN, S.; RAMAKRISHNA, S. V. Engineering aspects of solid-state fermentation. Enzyme Microb Tech 1985; 7(6): 258-265.

[23]   CASTILHO, L. R. Economic analysis of lipase production by *Penicillium restrictum* in solid-state and submerged fermentations. Biochem Eng J 2000; 4(3): 239-247.

[24]   BEISSON, F.; TISS, A.; RIVIÈRE, C.; VERGER, R. Methods for lipase detection and assay: a critical review. Eur J Lipid Sci Tech 2000; 102(2): 133–153.

[25]   HASAN, F.; SHAH, A. A.; HAMEED, A. Methods for detection and characterization of lipases: A comprehensive review. Biotechnol Adv 2009; 27(6): 782-98.

[26]   PLOU, F. J.; FERRER, M.; NUERO, O. M.; CALVO, M. V.; ALCALDE, M.; REYES, F.; *et al.* Analysis of Tween 80 as an esterase / lipase substrate for lipolytic activity assay. Biotechnol 1998; 12(3): 183-186.

[27]   GRIEBELER, N.; POLLONI, A. E.; REMONATTO, D.; ARBTER, F.; VARDANEGA, R.; CECHET, J. L.; *et al.* Isolation and Screening of Lipase-Producing Fungi with Hydrolytic Activity. Food Bioprocess Tech 2009; 4(4): 578-586.

[28]   FREIRE, D. M. G. Seleção de microrganismos lipolíticos e estudo da produção de lipase por *Penicillium restrictum*. PhD Thesis. 1996, 174p.

[29]   ABD RAHMAN, R. N. Z. R.; LEOW, T. C.; SALLEH, A. B.; BASRI, M. *Geobacillus zalihae* sp. nov., a thermophilic lipolytic bacterium isolated from palm oil mill effluent in Malaysia. BMC Microbiol 2007; 7(1): 77-86.

[30]   KOUKER, G.; JAEGER, K-E. Specific and sensitive plate assay for bacterial lipases. Appl Environ Microb 1987; 53(1): 211-213.

[31]   FERNANDES, M. L. M. Produção de lipases por fermentação em estado sólido e sua utilização em biocatálise. PhD Thesis. 2007; 131p.

[32]   MOREIRA, C. G.; DAMASO, M. C. T.; VALADÃO, R. C.; COURI, S. Screening of lipolytic filamentous fungi and study of lipase production using three different reactors. In: 5th International Technical Symposium on Food Processing, Monitoring Technology in Bioprocesses and Food Quality Management, Potsdam: Germany 2009. pp. 892-895.

[33]   MACIEL, V. F. A, PACHECO, T. F.; GONÇALVES, S. B. Padronização do uso de corante rodamina B para avaliação de atividade lipolítica em estirpes fúngicas. Comunicado Técnico. Brasília: Embrapa Agroenergy; 2010.

[34]   MAFAKHER, L.; MIRBAGHERI, M.; DARVISHI, F.; NAHVI, I.; ZARKESH-ESFAHANI, H.; EMTIAZI, G. Isolation of lipase and citric acid producing yeasts from agro-industrial wastewater. New Biotechnol 2010; 27(4): 337-340.

[35] KWON S-JOON; BAEK D-HEOUN; LEE S-GOO; SUNG M-HEE. Simple and rapid screening method for microbial D-stereospecific peptidase and esterase. Biotechnol Tech. 1999; 13(10): 653-655.

[36] SANDOVAL, G, Marty a. Screening methods for synthetic activity of lipases. Enzyme Microb Tech 2007; 40(3): 390-393.

[37] KONARZYCKA-BESSLER, M.; BORNSCHEUER, U. T. A High-Throughput-Screening Method for Determining the Synthetic Activity of Hydrolases. Angew Chem Int Edit 2003; 42(12): 1418-1420.

[38] SCHMIDT, M,; BORNSCHEUER, U. T. High-throughput assays for lipases and esterases. Biomol Eng 2005; 22(1-3): 51-56.

[39] PEREIRA EB, CASTRO HF, MORAES FF, ZANIN GM. Kinetic Studies of Lipase from *Candida rugosa*. Appl Biochem Biotech 2001; 91-93: 739-752.

[40] GOMBERT, A. K.; PINTO, A. L.; CASTILHO, L. R.; FREIRE, D. M. G. Lipase production by Penicillium restrictum in solid-state fermentation using babassu oil cake as substrate. *Process Biochem* 1999; 35(1-2): 85–90.

[41] LOWRY, O. H.; ROSEBROUGH, N. J.; FARR, A. L.; RANDALL, R. J. Protein measurement with the folin phenol reagent. J Biol Chem 1951; 193(1): 265-75.

[42] BRADFORD, M. M. A rapid and sensitive method for the quantitation of microgram quantities of protein utilizing the principle of protein-dye binding. Anal Biochem 1976; 72(1-2): 248-254.

[43] SOARES, C. M. F.; CASTRO, H. F.; MORAES, F. F.; ZANIN, G. M. Characterization and Utilization of *Candida rugosa* Lipase Immobilized on Controlled Pore Silica. Appl Biochem Biotech 1999; 77-79: 745-757.

[44] PINSIRODOM, P.; PARKIN, K. L. Lipolytic Enzymes. In: WROLSTAD, R. E.; ACREE, T. E.; DECKER, E. A.; PENNER, M. H.; REID, D. S.; SCHWARTZ, S. J.; *et al.*, Eds. Handbook of Food Analytical Chemistry. Hoboken, Wiley-Interscience, 2004; pp. 369-384.

[45] LOWRY, R. R.; TINSLEY, I. J. Rapid Colorimetric Determination of Free Fatty Acids. J Am Oil Chem Soc 1976; 53(7): 470-472.

[46] SALUM, T. F. C. Produçao e Imobilizaçao de Lipase de *Burkholderia cepacia* LTEB11 para a Síntese de Ésteres Etílicos. PhD Thesis. 2010; 130p.

[47] KWON, D. Y.; RHEE, J. S. A simple and rapid colorimetric method for the determination of free fatty acids for lipase assay. J Am Oil Chem Soc 1986; 63(1): 89–92.

[48] STAMATIS, H.; XENAKIS, A.; MENGE, U.; KOLISIS, F. N. Kinetic Study of Lipase Catalyzed Esterification Reactions in Water-Oil Microemulsions. Biotechnol and Bioeng 1993; 42(8): 931 -937.

[49] WINKLER, U. K.; STUCKMANN, M. Glycogen, Hyaluronate, and Some Other Polysaccharides Greatly Enhance the Formation of Exolipase by *Serratia marcescenst*. J Bacteriol 1979; 138(3): 663-670.

[50] SALUM T. F. C., BARON A. M.; ZAGO, E.; TURRA, V.; BARATTI, J.; MITCHELL, D. A.; *et al.* An efficient system for catalyzing ester synthesis using a lipase from a newly isolated *Burkholderia cepacia* strain. Biocatal Biotransfor 2008; 26(3): 197-203.

[51] OLIVEIRA, D.; FEIHRMANN, A. C.; DARIVA, C.; CUNHA, A. G.; BEVILAQUA, J. V.; DESTAIN, J.; *et al.* Influence of compressed fluids treatment on the activity of *Yarrowia lipolytica* lipase. J Mol Catal B Enzym 2006; 39(1-4): 117-123.

[52] JAEGER, K. E.; DIJKSTRA, B. W.; REETZ, M. T. Bacterial Biocatalysts: Molecular Biology, Three-Dimensional Structures, and Biotechnological Applications of Lipases. Annu Rev Microbiol. 1999; 53: 315-351.

Send Orders for Reprints to reprints@benthamscience.net

Methods to Determine Enzymatic Activity, 2013, 195-207 195

CHAPTER 8

Assays of Phenoloxidase Activity

Priscilla F. F. Amaral* and Bernardo D. Ribeiro

School of Chemistry, Federal University of Rio de Janeiro, Rio de Janeiro, Brazil

Abstract: Phenoloxidases are enzymes with broad substrate specificity tward a lot of phenols with several biotechnological applications. There are an expressive amount of methods in literature available for determining their activities. However, spectofotometric methods are much more practical and with very good evaluation by the users. Kinetic studies show that methods that measure the appearance of a product are much better over methods that measure substrate disappearance. Therefore, those kinds of methods are shown here.

Kwywords: Phenoloxidase, phenolic compounds, polyphenols, diamines, anilines, tyrosinase, laccases, dopaquinone, *o*-quinones, diphenolase activity, p-Cresol, catechol, oxygen, L-Dopa, melanin, ascorbic acid, dopachrome, L-tirosine, spectrophotometer, Agaricus bispora.

INTRODUCTION

Phenoloxidases are copper enzymes that catalyze the oxidation of phenolic compounds, as substituted phenols, polyphenols, diamines and anilines, using oxygen atoms as electron acceptors. They can be applied in textile dye and pulp bleaching, modification of lignocellulosic materials, wine, beer and fruit juices stabilization, production of polyphenolic polymers, effluent detoxification, phenol biosensors and soil bioremediation. Furthermore, they can be subdivided in two major types: tyrosinases and laccases [1-6].

Tyrosinase or polyphenol oxidase (PPO) (monophenol, *o*diphenol: O_2 oxidoreductase, EC 1.14.18.1), is an oxygenase oxyreductase found in several life forms, including the mushroom *Agaricus bisporus*, with great potential as a biocatalyst for applications involving biomodification of phenols or bioremediation of phenol-polluted water [7]. This enzyme has a binuclear copper center that catalyzes two different reactions: the orthohydroxylation of

---

*Address correspondence to Priscilla F. F. Amaral: School of Chemistry at Federal University of Rio de Janeiro, Brazil; Tel: 0055-21-2562 7579; Fax: 0055-21-2562 7622; E-mail: pamaral@eq.ufrj.br

monophenols to *o*-diphenols (monophenolase activity) and the oxidation of *o*-diphenols to their *o*-quinones (diphenolase activity). *p*-Cresol and catechol have been most frequently employed as experimental substrates for studying the two types of oxidation. Consequently, the two activities of this enzyme have come to be known as "cresolase" and "catecholase" activities, respectively [8]. It requires molecular oxygen to convert monophenols in *o*-hydroquinones and quinones in *o*-benzoquinones. The benzoquinones undergo a nonenzymatic polymerization to yield water-insoluble substances [9]. The PPO catalytic site has a unique dinuclear copper centre, in which the dioxygen molecule is bound and becomes polarized, resulting in its cleavage. The reducing substrate is then oxidized by the oxygen–copper complex.

Laccases (benzenediol:oxygen oxidoreductase, EC 1.10.3.2) belong to a larger group of enzymes termed multicopper oxidase superfamily which includes the plant ascorbate oxidase, the mammalian and avian plasma protein ceruloplasmin and bilirubin oxidase, yeast ferroxidases and bacterial metal oxidases [9,10]. Laccases have been found in plants, insects, bacteria, but are predominant in fungi (mainly in Basidiomycetes, Ascomycetes and Deuteromycetes) [3,11,12]. Their functions are related to lignin biosynthesis and degradation, pigment production, pathogenic virulence, metal transport and homeostasis [10,13].

Laccases contain four copper atoms bound to three redox sites designated Type I (T1), Type II (T2) and Type III (T3), according to their spectroscopic and paramagnetic properties. They react oxidizing the reducing substrate by T1 Cu^{2+}-mediated abstraction of one electron, forming a free radical, which can further undergo laccase-catalysed oxidation (*e.g.,* phenol to quinone) or non-enzymatic reactions (*e.g.,* hydration or polymerization). Each electron extracted from the four monoelectronic oxidations at the T1 site is transferred to the trinuclear cluster where O_2 is bound. Thus the T2 and T3 sites are the locations where reduction of molecular oxygen and the release of water occur [3,9,14,15].

Based on the mechanism above presented some Enzymatic Activity Measurements will be presented:

Tyrosinase Assay

PPO is an enzyme with a wide range of substrate specificity, which has led to many methods being proposed to measure its activity: radiometric, electrometric, chronometric and, especially spectrophotometric, which are fast and affordable by most laboratories [16]. The spectrophotometric technique has attracted more attention mainly because it is convenient, sensitive, and inexpensive and allows the course of the reaction to be studied continuously [17].

The fact that PPO catalyzes the oxidation of its monophenol/ o-diphenol substrates, giving rise to unstable o-quinones, has led some authors to propose that the best way to measure the enzymatic activity is to measure the disappearance of the monophenol or o-diphenol [17]. However, when substrate disappearance is measured at ultraviolet wavelengths, the molar absorptivity coefficients are usually high and the concentration cannot be increased much to reach PPO saturation [16]. Garcia-Molina *et al.* [16] discussed several spectrofotometric methods and for kinetically characterizing a series of monophenols and diphenols acting as PPO substrates. They found out that the methods that measure the appearance of a product and its evolution to a stable chromatophoric compound or the disappearance of ascorbic acid as coupled reagent show clear advantages over methods that measure substrate disappearance. They have also shown that as long as there is a clear stoichiometry in the evolution of the product of an enzymatic reaction, the initial velocity of the enzyme can be estimated measuring the final stable compound.

Method 1

Several studies indicate that tyrosinase catalyzes the conversion of L-tyrosine to L-3,4-dihydrohyphenylalanine (L-dopa) and then to melanin by the following reactions:

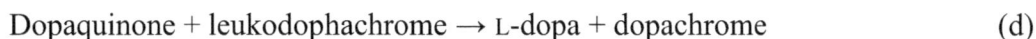

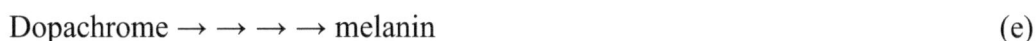

Tyrosine $\rightarrow$ L-dopa　　　　　　　　　　　　　　　　　　　　　　　　(a)

2 L-Dopa $\rightarrow$ 2 *O*-dopaquinone + 2 H_2O　　　　　　　　　　　　　　(b)

Dopaquinone $\rightarrow$ leukodophachrome　　　　　　　　　　　　　　　　(c)

Dopaquinone + leukodophachrome $\rightarrow$ L-dopa + dopachrome　　　　(d)

Dopachrome $\rightarrow \rightarrow \rightarrow \rightarrow$ melanin　　　　　　　　　　　　　　　　　(e)

The two initial reactions are enzymatically catalized by "creolase" and "catecholase", respectively. The catecholase activity of tyrosinase can be kinetically measured by the extent of dopachrome formation at 475 nm (using L-dopa as the substrate). A coupled assay can also be employed using the oxidation of ascorbic acid (measured at 265 nm) to determine the amount of dopaquinone formed from L-dopa [8].

In this method, the determination of catecholase activity of tyrosinase is done by two assay methods: measurement of dopachrome formation using L-dopa as substrate and measurement of dopaquinone formation from L-dopa with the oxidation of ascorbic acid.

Principle of the Method

Measuring dopachrome formation: the oxidation of L-dopa is followed by measuring the initial rate of increase in absorbance at 475 nm, at which dopachrome has a molar absorptivity of 3700 M^1.

Measuring dopaquinone formation: a fast nonenzymatic oxidation of ascorbic acid coupled to the reduction of dopaquinone enzymatically formed from L-dopa. Ascorbic acid is not oxidized by the enzyme. Therefore, the rate of ascorbate oxidation by dopaquinone, measured by the decrease in absorbance at 265 nm, is directly proportional to the extent of L-dopa oxidized to dopaquinone.

Procedure

Reagents

1. Tyrosinase extract;

2. Potassium phosphate monobasic;

3. Potassium dihydrogen phosphate;

4. L-dopa;

5. Ascorbic acid;

6. EDTA;

7.   Metaphosphoric acid.

Materials and Equipment

1.   Pipettes (10 μL, 100 μL, 1 mL and 5 mL);

2.   Glass flask (10 mL);

3.   1-cm oath length quartz cell;

4.   Photodiode array spectrophotometer.

Preparation of Solutions

Potassium Phosphate Buffer 100 mM pH 7.0

Dissolve 10.717 g of potassium phosphate monobasic and 5.242 g of potassium dihydrogen phosphate in 1000 mL distilled water to get 1000 mL of the buffer at the pH 7.0. Check in a pH meter and if the pH deviates from the desired value, slowly add 7 *M* phosphoric acid or 7M sodium hydroxide solution to reach pH 7.0.

Tyrosinase

Prepare a stock solution of 50 μg mL^{-1} in *potassium phosphate buffer 100 mM pH 7.0*.

Ascorbic Acid 1 mM

A stock solution is prepared dissolving 17.6 mg of ascorbic acid and 0.5 mg of EDTA in *potassium phosphate buffer 100 mM pH 7.0* containing 0.1% (w/v) metaphosphoric acid.

L-dopa 1 mM

Dissolve 19.7 mg of L-dopa in potassium phosphate buffer 100 mM pH 7.0.

Operation

Measuring dopachrome formation: add 1650μL of potassium phosphate buffer, pH 7.0 (0.1 *M*) and 1350 μl of L-dopa 1 m*M* in a glass flask and aerate this solution for 5 minutes. Then, add 2000 μL of this solution and 1000 μL of tyrosinase previously diluted to a quartz curvette. Immediately, the reaction progress must be followed in a spectrophotometer, at 280 nm for 60 to 90 s.

Measuring dopaquinone formation: add 1140 μL of potassium phosphate buffer, pH 7.0 (0.1 *M*), 1350 μL of L-dopa 1 m*M* and 510 μL of ascorbic acid 1 m*M* in a glass flask and aerate this solution for 5 minutes. Then, add 2000 μL of this solution and 1000 μL of tyrosinase previously diluted to a quartz curvette. Immediately, the reaction progress must be followed in a spectrophotometer, at 280 nm for 1 to 2 min.

For both measurements control experiments (blank runs) were always carried out using the same procedure, but substituting enzyme preparation for phosphate buffer.

Results and Conclusions

The maximal rate of L-dopa oxidation, measured by the extent of dopaquinone formation, is about twice the maximal rate of formation of dopachrome from L-dopa. This is in accord with the stoichiometry equation of 2 dopaquinone → dopachrome + L-dopa from the sum of reactions (c) and (d).

The two assays are complementary and yield the same value of K_M when L-dopa is used as the substrate. Assays at the K_M concentration of L-dopa (0.3 m*M*) can be used in conjunction with either method to estimate the catecholase activity of a tyrosinase solution.

Method 2

Principle of the Method

Polyphenol oxidase oxidizes tyrosine to dihydroxyphenylalanine which in turn is oxidized to o-quinone. The latter is accompanied by an increase in absorbance at 280 nm. The rate of increase is proportional to enzyme concentration and linear during a period of 5-10 minutes after an initial lag. One unit causes a change in absorbance at 280 nm of 0.001 per minute at 25°C, pH 6.5 under the specified conditions.

Procedure

Reagents

1. Tyrosinase extract;

2. Potassium phosphate monobasic;

3. Potassium dihydrogen phosphate;

4. L-tyrosine.

Materials and Equipment

1. Pipettes (10 µL, 100 µL, 1 mL and 5 mL);

2. Glass flask (10 mL);

3. 1-cm oath length quartz cell;

4. UV spectrophotometer.

Preparation of Solutions

Potassium Phosphate Buffer 50 mM pH 6.5

Dissolve 3.136 g of potassium phosphate monobasic and 4.359 g of potassium dihydrogen phosphate in 1000 mL distilled water to get 1000 mL of the buffer at the pH 6.5. Check in a pH meter and if the pH deviates from the desired value, slowly add 7 M phosphoric acid or 7 M sodium hydroxide solution to reach pH 6.5.

Tyrosinase

Dissolve at a concentration of 1 mg/mL in distillate water. Dilute further in reagent grade water to a concentration of 200-400 units mL^{-1}.

L-tyrosine 1 mM

Dissolve 18.1 mg of L-tyrosine in 100 mL of distillate water and agitate vigorously until complete dissolution.

Operation

Add 2.5 mL of L-*tyrosine* 1 m*M* and 2.5 mL of p*otassium phosphate buffer 50 mM pH 6.5* in a glass flask and aerate this reaction mixture by bubbling oxygen for 5 minutes. Then, add 2900 μl of this mixture to a quartz curvette. Transfer cuvette to the spectrophotometer and record the absorbance at 280 nm (A_{280}) for 4-5 minutes to achieve temperature equilibration and to establish blank rate, if any. Add 0.1 mL of appropriately diluted enzyme and record A_{280} for 10-12 minutes. Determine ΔA_{280} from the linear portion of the curve. A non-linear "lag" of 2-3 minutes can be expected.

Calculation of Tyrosinase Activity

$$Units/ml = \frac{\Delta A_{280}/min * 1000 * D_E}{V_E} \qquad (1)$$

Where:

ΔA_{280}/min is the absorbance variation in a linear increase per minute,

D_E is the dilution of the enzyme extract and

V_E is the volume of enzyme used in the assay (0.1 mL).

Results and Conclusions

This method shows excellent results to be used for mushroom (*Agaricus bispora*) tyrosinase extract [7, 19].

Perspectives

Despite the variety of methods to determine tyrosinase activity, none could be considered "perfect". Measurements of oxygen consumption allows PPO activity to be expressed in katals as is recommended by IUPAC; however, lack of specificity is a well-known drawback to this type of assay [20]. In spectofotometric methods, the practicality and good results justify its use by several laboratories. Kinetic studies [16] prove that methods that measure the appearance of a product, which are shown here, are much better over methods that measure substrate disappearance.

Laccase Assay

Method 1

Laccase's primary substrates may be 2,2'-azino-bis(3-ethylbenzthiazoline-6-sulfonic acid) [ABTS] or 1-hydroxybenzothiazole [HBT] and in the presence of those substrates it can catalyze the oxidation of nonnatural substrates, including polycyclic aromatic hydrocarbons. The catalytic capabilities of this enzyme can be greatly enhanced by the addition of suitable mediator compounds. The mechanism of oxidation by laccase-mediator systems is still under discussion [21]. The method shown here is based on the oxidation of ABTS and the concomitant formation of a radical cation.

Principle of the Method

The laccase activity is usually determined by a colorimetric method, based on the change of absorbance due to the formation of oxidation product, which depend or not of mediator as ABTS and HBT, as in Table **1** [21,23].

Table 1: Assays for Screening Laccase Activity

Analyte	Product	Wave length (nm)
ABTS (2,2'-azino-bis(3-ethylbenzthiazoline-6-sulfonic acid)	ABTS radical cation	418
Anthracene/Sodium borohydride	9,10-Anthrahydroquinone	419
Poly R-478/HBT (1-hydroxy benzothiazole)	Polymeric dye decolorized	520
Sodium Iodide/ABTS	Triiodide	353

Procedure

Reagents

1. Sodium dihydrogen phosphate monohydrate;

2. Sodium hydrogen phosphate heptahydrate;

3. Sodium hydroxide (pearls);

4. Phosphoric acid;

5. ABTS;

6. PEG (polyethylene glicol) 5000;

7. Enzyme extract;

8. Distilled water.

Materials and Equipment

1. Plastic microtubes (2mL, polypropylene);

2. Pipettes (10 μL, 100 μL, 1 mL and 5 mL);

3. Glass flask (100 mL);

4. Glass cuvette;

5. Spectrophotometer (visible range) coupled to a computer;

6. Thermostatic bath (preferably digital, 0.1 °C precision);

7. Vortex mixer.

Preparation of Solutions

Phosphate Buffer 100 mM pH 6.0

Dissolve 1.2143 g of sodium dihydrogen phosphate monohydrate and 0.3218 g of sodium hydrogen phosphate heptahydrate in 100 mL distilled water to get 100 mL of the buffer at the pH 6.0. Check in a pH meter and if the pH deviates from the desired value, slowly add 7 *M* phosphoric acid or 7 *M* sodium hydroxide solution to reach pH 6.0.

ABTS Solution 3.0 mM

For the preparation of ABTS solution, 1.544 g of ABTS was cautiously added to 50 mL of phosphate buffer 100 mM pH 6.0, and 10 mL PEG 5000 solution 50% w/v and then diluted to 100 mL with water. The solution should be stored refrigerated for no longer than 1 month in amber glass flasks, labeled with date of preparation as well as solution and operator identifications.

Operation

Assay of Enzyme Extracts (Arnold and Georgiou, 2003)

For the determination of laccase activity in enzyme extracts with unknown activities, add 2700 μL of the ABTS solution and 300 μL of enzyme preparation, previously diluted with phosphate buffer 100 mM pH 6.0, to a glass cuvette, at 25 ∘C. Immediately, the reaction progress must be followed in a spectrophotometer, at 418 nm, coupled to a computer, for 100 s. Control experiments (blank runs) were always carried out using the same procedure, but substituting enzyme preparation for phosphate buffer.

Calculation of Laccase Activity

For the calculation of activity, use Equation 2. The range of absorbance to be used should be corresponding to the initial velocity of enzymatic reaction, thus, where there is a linear relationship. One unit of laccase activity (LAC) is defined as representing the amount of enzyme enough to oxidize 1 μmol of ABTS per minute per milliliter, under the assay conditions.

$$POD\ (U\ mL^{-1}) = \frac{\Delta Abs}{\varepsilon \times l \times \Delta t \times V_{ENZ}} \qquad (2)$$

Where:

ε is the molar absorption coefficient of green ABTS radical cation (36000 $M^{-1}.cm^{-1}$);

l is the path length (1 cm);

Abs is the absorbance values registered for spectrophotometer;

t represents the time of enzymatic reaction (min); and

V_{ENZ} represents the volume of enzyme extract (mL) used for the assay.

RESULTS AND CONCLUSIONS

Laccase from several microorganisms could oxidize ABTS [12,22,24], but not all of them can oxidize HBT [22]. Therefore, the method described here shows a greater use to detect laccase activity.

ACKNOWLEDGEMENTS

The authors thank the financial support of the Brazilian National Council for Scientific and Technological Development (CNPq), (MCT- CNPq) and the Carlos Chagas Filho Foundation for Research Support in the State of Rio de Janeiro (FAPERJ).

CONFLICT OF INTEREST

The authors confirm that this chapter content has no conflict of interest.

REFERENCES

[1]   MINUSSI, R. C.; PASTORE, G. M.; DURÁN, N. Potential applications of laccase in the food industry Trends Food Sci. Technol., 2002, 13, 205–216.
[2]   DE FARIA, R. O.; MOURE, V. R.; AMAZONAS, M. A. L. A.; KRIEGER, N.; MITCHELL, D. A. The Biotechnological Potential of Mushroom Tyrosinases Food Technol. Biotechnol., 2007, 45 (3), 287-294.
[3]   POLAINA, J.; MACCABE, A. P. Industrial Enzymes: Structure, Function and Applications; Springer, Dordrecht, The Netherlands, 2007.
[4]   WIDSTEIN, P.; KANDELBAUER, A. Laccase: applications in the forest products Enzyme Microbial Technol., 2008, 42, 293–307.
[5]   CAÑAS, A. I.; CAMARERO, S. Laccases and their natural mediators: Biotechnological tools for sustainable eco-friendly processes Biotechnol. Adv., 2010, 28, 694–705.
[6]   KUDANGA, T.; NYANHONGO, G. S.; GUEBITZ, G. M.; BURTON, S. Potential applications of laccase-mediated coupling and grafting reactions: A review Enzyme Microbial Technol., 2011, 48, 195–208.
[7]   ATLOW, S. C.; BONADONNA-APARO, L. AND KLIBANOV, A. M. Dephenolization of industrial wastewaters catalyzed by polyphenol oxidase. Biotechnol. Bioeng., 1984, 26 (6), 599-603.

[8]   BEHBAHANI, I; MILLER, S. A.; O'KEEFEE, D. H. A comparison of Mushroom tyrosinase dopaquinone and dopachrome assays using diode-array spectrophotometry: dopachrome formation *vs.*ascorbate-linked dopaquinone reduction. Microchem. J., 1993, 47, 251-260.

[9]   SOLOMON, E. I.; SUNDARAM, U. M.; MACHONKIN, T. E. Multicopper oxidases and oxygenases. Chem. Rev., 1996, 96, 2563–2606.

[10]   MCCAIG, B. C.; MEAGHER, R. B.; DEAN, J. F. D. Gene structure and molecular analysis of the laccase-like multicopper oxidase (LMCO) gene family in *Arabidopsis thaliana*. Planta, 2005, 221, 619-636.

[11]   BALDRIAN, P. Fungal laccases: occurrence and properties FEMS Microbiol. Rev., 2006, 30, 215–242.

[12]   SHARMA, P.; GOEL, R.; CAPALASH, N. Bacterial laccases World. J. Microbiol. Biotechnol., 2007, 23, 823-832.

[13]   MAYER, A. M.; Staples, R.C. Laccase: new functions for an old enzyme Phytochem., 2002, 60, 551–565.

[14]   DURÁN, N.; ROSA, M. A.; D'ANNIBALE, A.; GIANFREDA, L. Applications of laccases and tyrosinases (phenoloxidases) immobilized on different supports: a review. Enzyme Microbial Technol., 2002, 31, 907–931.

[15]   SAKURAI, T.; KATAOKA, K. Basic and Applied Features of Multicopper Oxidases, CueO, Bilirubin Oxidase, and Laccase. The Chem. Rec., 2007, 7, 220–229.

[16]   GARCÍA-MOLINA, F.; MUÑOZ, J. L.; VARÓN, R.; RODRÍGUEZ-LÓPES, J. N.; GARCÍA-CÁNOVAS, F.; TUDELA J. A Review on Spectrophotometric Methods for Measuring the Monophenolase and Diphenolase Activities of Tyrosinase. J. Agric. Food Chem., 2007, 55, 9739–9749.

[17]   HAGHBEENA, K.; TANB, E. W. Direct spectrophotometric assay of monooxygenase and oxidase activities of mushroom tyrosinase in the presence of synthetic and natural substrates. Anal. Biochem., 2003, 312, 23–32.

[18]   Worthington, C.C. Worthington Enzyme Manual. Worthington Biochemical Corporation, Freehold, NJ, 1988, pp. 74-75.

[19]   AMARAL, P. F. F.; FERNANDES, D. L. A.; TAVARES, A. P. M.; XAVIER, A. M. R. B.; CAMMAROTA, M. C.; COUTINHO, J. A. P.; COELHO, M. A. Z. Decolorization of dyes from textile wastewater by *Trametes versicolor* Environ. Technol., 2004, 25, 1313-1320.

[20]   GAUILLARD, F.; RICHARD-FORGET, F.; NICOLAS J. New spectrophotometric assay for polyphenol oxidase activity. Anal. Biochem., 1993, 215, 59–65.

[21]   ARNOLD, F. H.; Georgiou, G. Directed Enzyme Evolution: Screening and Selection Methods, Methods in Molecular Biology. Vol. 230; Humana Press, Totowa, New Jersey, 2003.

[22]   ANDER, P.; MESSNER, K. Oxidation of 1-hydroxybenzotriazole by laccase and lignin peroxidase. Biotechnol. Tech., 1998, 12 (3), 191–195.

[23]   REYMOND, J.-L. Enzyme Assays: High-throughput Screening, Genetic Selection and Fingerprinting; WILEY-VCH Verlag GmbH & Co. KGaA, Weinheim, Germany, 2006.

[24]   TAVARES, A.P.M.; COELHO, M. A. Z.; COUTINHO, J. A. P.; XAVIER, A. M. R. B. Laccase improvement in submerged cultivation: induced production and kinetic modeling J. Chem. Technol. Biotechnol. 2005, 80, 669–676.

Send Orders for Reprints to reprints@benthamscience.net

CHAPTER 9

Assay Method for Transglutaminase Activity

Mônica Caramez Triches Damaso[1,*], Romulo Cardoso Valadão[2], Sonia Couri[3] and Alane Beatriz Vermelho[4]

[1]*Embrapa Agroenergy, Brasília, and Embrapa Food Technology, Rio de Janeiro, Brazil;* [2]*Federal Rural University of Rio de Janeiro, Institute of Technology, Seropédica, Rio de Janeiro, Brazil;* [3]*Federal Institute of Education, Science and Technology of Rio de Janeiro, Rio de Janeiro, Brazil and* [4]*Department of General Microbiology, Institute of Microbiology Paulo de Góes- Biotechnology Center-Bioinovar, Federal University of Rio de Janeiro, Rio de Janeiro, Brazil*

Abstract: Transglutaminases are enzymes capable of catalyzing acyl transfer reactions by introducing covalent cross-links between proteins, as well as peptides and various primary amines. These enzymes are very important due to their abilities to modify proteins and act as biological glues. Nowadays, microorganisms are considered the best source of transglutaminases, due to the difficulties involved in producing them from mammal tissues an industrial scale. In this chapter is described an analytical method for the transglutaminase (TGase) activity assay. Furthermore, the qualitative and quantitative methods for screening microorganisms that produce this enzyme will also be detailed

Keywords: Aminoacyl transferases, transglutaminase, γ-carboxyamide group, food processing, blood clotting, amide groups, glutamine, hydroxylamine, body fluids, tissues animals, microorganisms, *streptoverticillium, streptomyces*.

INTRODUCTION

Transglutaminase (TGase, EC 2.3.2.13) is an enzyme that belongs to the group of aminoacyl transferases and according to the Enzyme Commission it is a protein-glutamine γ-glutamyltransferase.

TGase catalyzes the acyl transfer reactions between γ-carboxyamide groups of glutamine residues in proteins, peptides and primary amines. During these reactions the γ-carboxyamide group act as acyl donors, while the acyl acceptors

---

*Address correspondence to Mônica Caramez Triches Damaso: Embrapa Agroenergy, Brasília, and Embrapa Food Technology, Rio de Janeiro, Brazil; Tel: + 55 (61) 3448 2328; Fax: + 55 (61) 3448 1589; E-mail: monica.damaso@embrapa.br

can be primary amines or ε-amino groups of lysine residues either as free lysine or as peptide/protein bound. In the last case, both intra and inter molecularly bonds are formed. In the absence of these acceptor groups water can act as acceptor, occurring the deamidation of the γ-carboxyamide groups, transforming glutamine residues into glutamic acid [1, 2]. Then, TGase can modify proteins by the introduction of intra and inter molecularly cross-links, as well as through the incorporation of amines and by deamidation (Fig. **1**).

Heinrich B. Waelsch discovered in 1957, the occurrence of these enzymes [3] and since then, studies dealing with its production, purification, properties and applications have been exhaustively performed.

The TGase are found in various organisms ranging from bacteria to mammals [4], including plants, microorganisms and body fluids and tissues of various animals [5].

TGases are involved in several biological processes, including blood clotting, wound healing, epidermal keratinization and stiffening of the erythrocyte membrane [6].

The enzyme of animal origin, obtained from the Guinea pig (*Cavia aperea*), was the sole source of TGase production for decades. The shortage of supply, the difficult process of separation and purification, added to the fact that they are calcium dependent and the extremely high price of the product led to recent efforts to produce TGase by microorganisms [7]. TGase extracted from animal tissue are calcium-dependent, whereas TGase produced by microorganisms does not require calcium ions to promote their reactions, and are stable over long storage time. These properties increased the agroindustrial interest in this kind of enzyme [2, 8, 9].

The literature reports two microbial sources as TGase producers. The first research involving microbial sources was described by Ando *et al.* [1] that reported the screening of 5000 strains isolated from Japan soil. The strongest TGase producer was an actinomycete strain of *Streptoverticillium* S-8112. Thenceforth, other strains were isolated, such as *Streptoverticillium mobaraense* and *Streptoverticillium griseocarneum* [7], *Streptomyces hygroscopicus* [10] and *Streptomyces platensis* [11]. The other microbial groups that produce TGase are bacilli strains of *Bacillus subtilis* [12] and *Bacillus circulans* [13-15]. No wild yeast or filamentous fungi were identified as TGase producer to date. The enzyme from *B. subtilis* catalyses the cross-linking of the spore coat protein GerQ thereby contributing to the physical and chemical resistance of the bacterial spore [16].

The use of genetically modified microorganisms in the TGase production [4, 17, 18] has been motivated in order to produce high amounts of enzymes at lower prices. However, there is still some degree of difficulty in the commercialization of these enzymes due to factors related to food regulation and consumer acceptability.

TGase can be used in different applications: textiles and leather [19, 20], pharmaceuticals [21] and mainly food processing [22].

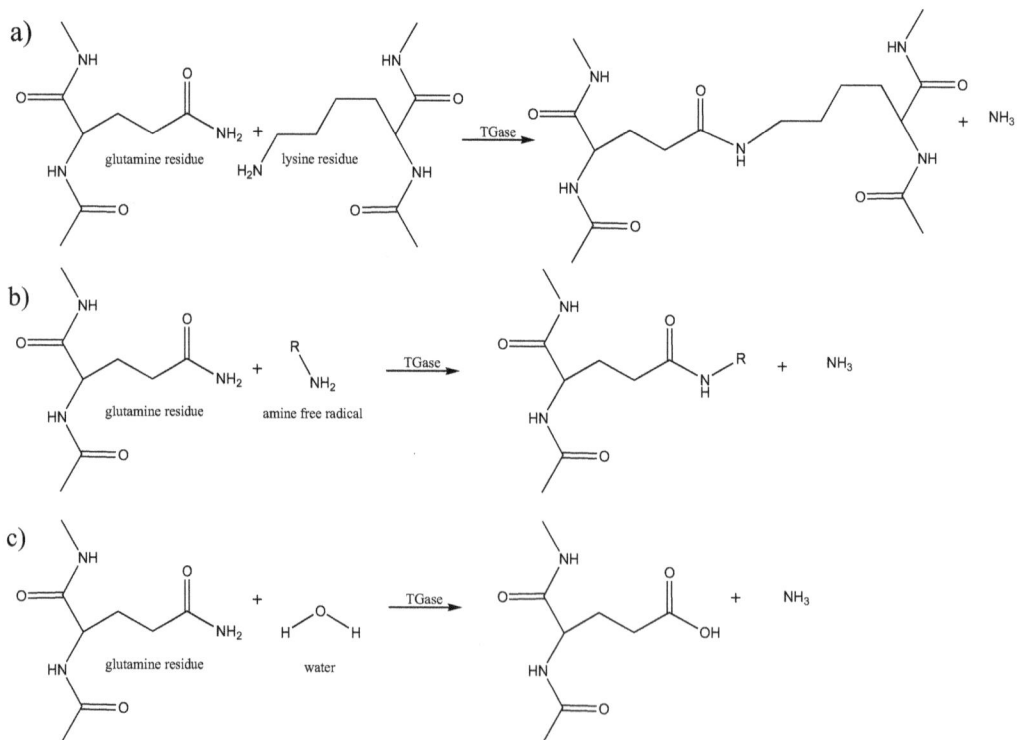

Figure 1: Reactions catalyzed by TGase: a) cross-link between glutamine and lysine of protein or of peptidic chain; b) incorporation famine c) deamidation.

The enzyme catalyzes physical changes in the food structure that increase its aggregate value [23]. TGase action improve water-holding capacity, elasticity and appearance, can promote protein texturization and gelation, modify solubility and nutritional value [22]. The effects of these changes increased the interest in different food industrial sectors: poultry [24], bovine [23], fish [5], dairy [25], bakery and dough [26; 27]. Furthermore, the TGase-mediated reactions generate products with the ability to form gels, with new textures and viscosities, beyond increasing physical resistance and thermal stability of certain foods [5, 22-24].

The application of commercial TGase on the elaboration of salmon medallions (Fig. **2**) is an example of the ability of the enzyme to aggregate the trimmings generated in the filleting of salmon [28].

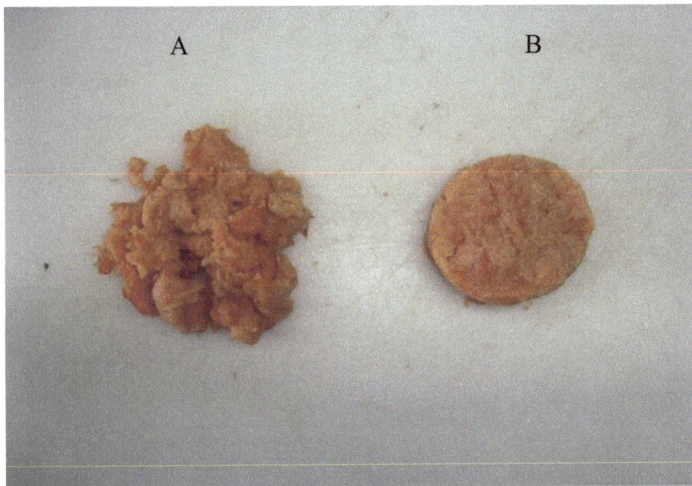

Figure 2: Salmon trimmings (A) and medallion (B) obtained by the action of TGase.

Due to the recent developments on the production, purification, characterization and application of this enzyme, there is a need for standardized methods in order to compare the catalytic efficiency of TGases obtained from different sources in distinct laboratories.

There are several methods available for measuring TGase activity and they follow different principles of quantification. Some of them are direct methods and measure the enzyme activity based on the disappearance of amino groups or incorporation of amines or formation of NH_3. On the other hand, indirect methods are able to evaluate the enzyme action through the amount of protein product obtained or measurement of the functional effects as viscosity or gel strength [2].

These methods can use spectroscopy [29, 30, 31], electrophoresis [32-34], viscosimetry [32], chromatography techniques [34, 35] and scanning electronic microscopy [33].

As the focus of this chapter is to evaluate the enzyme activity, a colorimetric hydroxamate procedure that is the most common assay method used for microbial TGase will be described. Furthermore, the qualitative and quantitative methods for screening microorganisms that produce TGase will also be detailed.

Screening for Microorganism that Produce Transglutaminase

Screening for microorganisms that produce TGase cannot be accomplished through the use of a selective agar medium as often performed for screening cellulase [36] and lipase [37] producing microrganisms. Then, in the last two decades the screening procedure applied to TGase producers is carried out in two joint steps: 1) the production of enzyme in a small-scale, usually by submerged fermentation and 2) analytical assay to determine enzyme activity (Qualitative method). Although this approach is still being used today, it is labor-intensive and time consuming. Thence, Yokoyama *et al.* [38] adapted a plate screening method using the same principle and reagents used in the TGase assay method (Qualitative method) [29, 30]. In the following text an example of both strategies will be described.

Detection of Microorganisms that Produce TGase by the Formation of Color Zones (Qualitative Method)

This method is based on the capacity of microorganisms to produce TGase. It is an adaptation of the colorimetric hydroxamate procedure used in the analytical assay of the enzyme activity [29, 30]. The TGase producing microorganisms can be identified by the red color formed around the colonies. On the other hand, the non-TGase producing strains show a yellow color [38].

Materials

1. Strains of interest;

2. Agar medium (*e.g.*: Yokoyama *et al.* [38]);

3. Solution 1: 0.12 M CBZ-Gln-Gly (Sigma-Aldrich C6154), 0.4 M hydroxylamine, 0.04 M glutathione (reduced form) and 0.2 M MES buffer, adjusted to pH 6.0 [38];

4. Solution 2: 1 N HCl, 4% TCA and 5% $FeCl_3.6H_2O$ [38];

5. Nitrocellulose membrane (ADVANTEC MFS, INC.);

6. Stainless steel plate.

Equipment

1. Analytical balance;

2. pHmeter;

3. Autoclave;

4. Incubation chamber.

Procedure

1. Transferred the strains of interest onto a nitrocellulose membrane;

2. Prepare the agar medium. After autoclaving, cool it to about 60°C. Distribute the agar into the petri dishes;

3. Incubate the membrane on the agar medium plate, in adequate conditions. This test should preferably be carried out using triplicates;

    e.g.: *Corynebacterium glutamicum* ATTC13869 containing a MTGase from *Streptomyces mobaraensis* was incubated at 30 °C for 24 hours [38].

4. After incubation, this membrane is transferred onto the stainless steel plate;

5.   Spray 6 mL of substrate solution 1 on the membrane and incubate it at 37°C for 45 min;

6.   Stop the reaction spraying on solution 2;

7.   Red color around the colony indicates positive strains while yellow indicates the negative ones.

Comments

Among the studies described in the literature, Yokohama *et al.* [38] were the first to propose a qualitative method for screening microorganisms that produce TGase. The agar medium used by Yokoyama *et al.* [38] is complex and demanding, but it was necessary because the genetic manipulation applied to the microorganisms tested. However, it is believed that less complex media as the one described by Ando *et al.* [1] and used by others authors can also be applied.

Detection of Microorganisms that Produce TGase by Measuring the Enzyme Activity (Quantitative Method)

This method is based on the capacity of microorganisms to produce TGase. It is a method divided in two steps, the first, described in this topic, involves the enzyme production while the second one consists of the quantification of the enzyme activity value.

Materials

1.   Strain of interest;

2.   Sugar source;

3.   Nitrogen source;

4.   Salts;

5.   Sodium hydroxide or hydrochloric acid for pH adjust;

6.   Beakers;

7. Conical flasks or Falcon Tubes;

8. Platinum inoculating loop.

Equipment

1. Analytical balance;

2. pH meter;

3. Autoclave;

4. Incubation chamber;

5. Rotatory shaker.

Procedure

1. Prepare the medium. After autoclaving, cool it to a room temperature. Distribute the suitable medium amount into the conical flasks or Falcon tubes.

    e.g., (%wv^{-1} in distilled water): polypeptone (2.0), yeast extract (0.2), soluble starch (2.0), dipotassium hydrogen phosphate (0.2), magnesium sulphate (0.1) and foam-extinghisher (0.05). The pH of the medium is adjusted to 7.0 [1].

2. Inoculate microorganisms into the sterilized medium using a small amount of the strain previously grown. Another option is to use the pre-inoculum obtained after a 24h incubation of the strain in the same medium described above [15]. This test should preferably be carried out using duplicates.

    e.g.: Actinomycete strains were incubated 3 days at 30 °C and 250 rpm [1] and bacilli strains during 20 days at 30 °C and 180 rpm [39].

    The crude extract can be obtained after the incubation period using vacuum filtration and the TGase activity is measured by the quantitative method described as follows. The best strains will be those that produce the higher levels of enzymatic activity.

Transglutaminase Activity Assay

As described above, eventhough some methods can be used to determine the transglutaminase activity, the colorimetric hydroxamate one is usually used to quantitatively determine the enzyme activity.

This method is based on the formation of hydroxamic acids by the enzyme-catalyzed replacement of the amide groups of glutamine with hydroxylamine [29].

Materials

1. Transglutaminase, *e.g.,* crude extract from animal or microbial or commercial enzyme;

2. Tris-acetate;

3. Hydroxylamine;

4. Calcium chloride (only for animals TGase);

5. Reduced glutathione;

6. N-carbobenzoxy-L-glutaminyl-glycine (substrate) (CBZ-Gln-GlySigma-AldrichC6154);

7. Hydrochloric acid;

8. Trichloroacetic acid (TCA);

9. Ferric chloride hexahydrated;

10. L-glutamic acid-γ-monohydroxamate (standard) (Sigma-Aldrich G2253);

11. Volumetric flask;

12. Microfuge tubes of 1.5 or 2.0 mL;

13. Microcuvettes.

Equipment

1.   Analytical balance;

2.   pH meter;

3.   Water bath with temperature controller;

4.   Spectrophotometer suitable for measuring absorbance at 525 nm.

Procedure

Reagents and Solution [29, 30]

Reagent A

Dissolve 2.4228 g of buffer Tris-acetate (200mM), 0.6949 g of hydroxylamine (100 mM), 0.5549 g calcium chloride (50 mM) (only from animal enzyme), 0.3073 g reduced glutathione (10 mM) and 1.0119 g of CBZ-Gln-Gly (30 mM) in 100 mL of distilled water. Adjust to pH 6.0 using a 0.1M hydrochloric acid solution. The Reagent A should be prepared at the day of the analyses.

Reagent B

Solution of 1:1:1 (v:v:v) 3 N Hydrochloric acid, 12 % trichloroacetic acid; 5% ferric chloride hexahydrated dissolved in 0.1N hydrochloric acid. Store the solution in a bottle at 4°C for up to 1 month.

3N Hydrochloric Acid Solution

The concentration and the density of hydrochloric acid in the reagent bottle need to be considered for preparing this solution. Considering the acid concentration of 37% v/v, *i.e.,* 37 mL/100 mL and the density of 1.19 g/mL, put 249 mL of acid into volumetric flask containing distilled water. Mix and bring the volume to 1L with the addition of water.

12% Trichloroacetic Acid (TCA) Solution

   Dissolve 12.0 g of trichloroacetic acid in 50 mL of distilled water. Bring the volume to 100 mL with distilled water in a volumetric flask.

5% Ferric Chloride Hexahydrated Solution in 0.1N Hydrochloric Acid

   Dissolve 5g of ferric chloride hexahydrated in 50 mL of 0.1N hydrochloric acid and bring the volume to 100 mL with the acid.

Standard Solution of L-Glutamic Acid γ-Monohydroxamate (10 mM)

   Dissolve 0.1621 g of L-glutamic acid γ-monohydroxamate in 50 mL of distilled water. Bring the volume to 100 mL with the addition of water.

Enzymatic Assay

1. Place 400 µL of reagent A in microfuge tube;

2. Acclimate to 37°C;

3. Add 400 µL of enzyme (diluted if necessary) and mix;

4. Keep it incubated for 10 min at 37 °C;

5. Add 400 µL of reagent B to stop the reaction and mix;

6. Centrifuge at 10,000 rpm for 3 min. Use the supernatant (enzyme sample);

7. Zero the spectrophotometer at 525 nm with the zero point (Table **1**) of standard curve;

8. Measure the absorbance (A_{525}) of the supernatant (enzyme sample) at 525 nm;

9. Correct the absorbance obtained with the enzyme blank [1];

10. Using the standard curve [2] (Fig. **4**), convert the corrected absorbance to enzyme activity units (U);

11. Calculate the enzyme activity (U.mL^{-1}) using the Equation 1:

$$U.mL^{-1} = \frac{\left[\left(A_{ES} - A_{EB}\right) - b\right] \times D}{\left(a \times \Delta t\right)} \tag{1}$$

a = angular coefficient of the standard linear regression
A_{ES} = mean of absorbance of the enzyme sample
A_{EB} = mean of absorbance of enzyme blank
b = linear coefficient of the standard linear regression
D = enzymatic dilution if is necessary
Δt = incubation time (enzyme-substrate reaction).

One enzymatic unit (U) of TGase catalyzes the formation of 1 μmol hydroxamic acid per min of reaction at 37°C. Enzyme can be expressed as either volumetric activity (U.mL^{-1}) (Equation 1) or specific activity (U.g^{-1} protein).

Obs.: This test should preferably be carried out using triplicates.

[1] Enzyme blank sample:

Repeat the steps 1 to 6 described above in the enzymatic assay, however replacing the enzyme by a denatured one. Measure the supernatant absorbance at 525 nm.

Obs.: This reaction subtracts any interference that can be present in the enzyme or in the compounds of the reagents or caused by the reaction conditions that can modify the determination of the correct value of activity.

[2] L-glutamic acid-γ-monohydroxamate standard curve

The standard used in the assay is L-glutamic acid-γ-monohydroxamate (Sigma-Aldrich), prepared as described above. This solution can be stored at -20 °C. The standard solution can be diluted according to Table **1**, however, if necessary, more dilutions can be added.

Tubes containing the different concentrations of the standard solution are submitted to the procedure as follows:

Standard Curve Procedure

1. Place into the microfuge tube the amount of L-glutamic acid-γ-monohydroxamate and water corresponding to each point of the standard curve (Table **1**);

Table 1: Dilutions and respective concentrations of L-glutamic acid-γ-monohydroxamate used in standard curve

Acid Concentration (mM)	Acid Volume (µl)	Water Volume (µl)
0.00	0	400
0.25	10	390
0.50	20	380
0.75	30	370
1.00	40	360
1.50	60	340
2.00	80	320
2.50	100	300
3.00	120	280
3.50	140	260

2. Add 400 µL of Tris HCl Buffer pH 6.0 (Reagent A without substrate, hydroxylamine, calcium chloride and glutathione) and mix;

3. Add 400 µL of Reagent B and mix;

4. Centrifuge at 10,000 rpm for 3 min. Use the supernatant;

5. Measure the absorbance of each supernatant at 525 nm.

As a result, different color intensities are generated by the different acid concentrations, as shown in Fig. **3**, and the standard curve is constructed using the different values of acid concentration (X axis) *vs.* the correspondent absorbance (Y axis) (Fig. **4**).

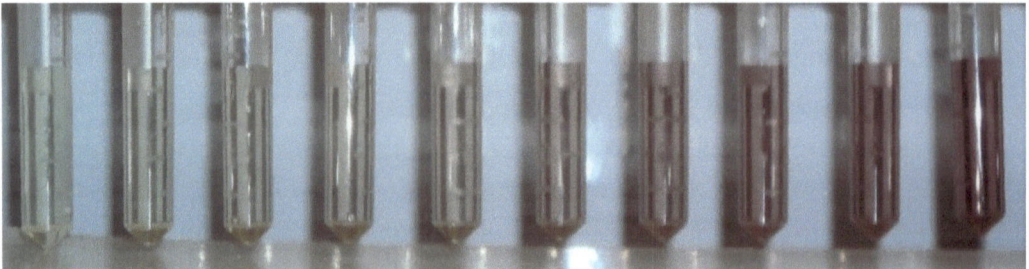

Figure 3: L-glutamic acid-γ-monohydroxamate standard showing the variation of colors generated by the different concentrations of acid.

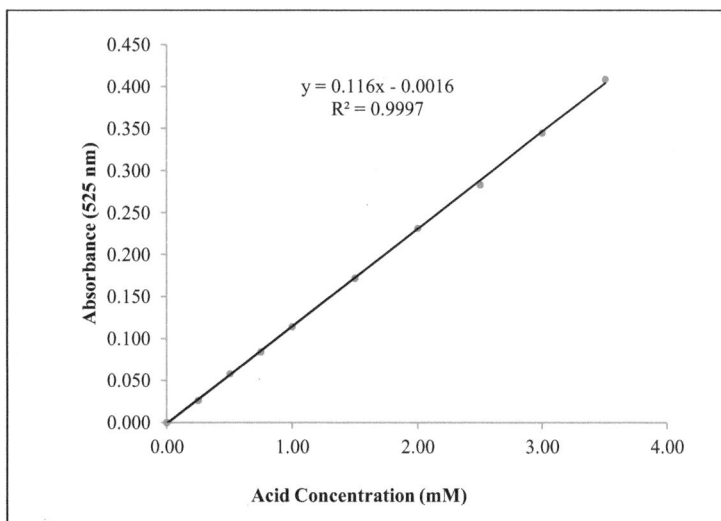

Figure 4: Acid standard curve indicating the angular and linear coefficients of the standard linear regression.

It is interesting to note that the standard curve can be prepared using concentration units (molarity, as shown in Fig. **4**) or amount units (micromoles, for instance). If the standard curve is prepared using concentration the calculation of the enzymatic activity is as specified in Equation 1. However, if it is used acid micromoles units the calculation of the original activity must be divided to 0.4 microliters which is the amount of enzyme used in the enzyme test.

Obs.: The standard curve should preferably be carried out using triplicates.

Comments

Several methods have been described in literature for the measurement of TGase activity using different types of strategies for the assays: i) the fluorometric or radioactive compounds and ii) the enzyme-linked colorimetric or immunochemical assays [40]. The various methods developed to measure TGase activity may present variations in the technique and different types of substrates are available. They present advantages and disadvantages and the best choice depends of the objectives and the type of TGase. This chapter described an analytical method assay to determine TGase activity, a colorimetric method which is based on hydroxamate to detect and quantify the transglutaminases. This method is one of the most commonly used methods for these enzymes and is easy, fast and sensitive. A sensitivity method with fluorescent substances also is very applicable. This method is based on the measurement of the rate of fluorescence enhancement due to the incorporation of fluorescent polyamines into protein substrates at λ_{exc} 260 nm and λ_{em} 538 nm [41-43]. The methods described by Jeitner *et al.* [31] and Jeitner *et al.* [43] are highly sensitive fluorometric plate assays and are suitable for high-throughput screening for transglutaminase activity. Both methods can be applied to any kind of TGase, although the first one is the most cited for microbial enzyme and the second for animal sources. Apart from the methods, Ca^{2+} needs to be added for the success of animal enzyme determination, while it is not necessary for microbial one.

Equipment

1.   Analytical balance;

2.   pHmeter;

3.   Autoclave;

4.   Incubation chamber;

5.   Rotatory shaker;

6.   Water bath with temperature controller;

7.   Spectrophotometer.

ACKNOWLEDGEMENTS

The authors thank Embrapa, UFRRJ, FAPERJ and MCT-CNPq (all from Brazil) for the financial support and the researcher Admilson Costa da Cunha who kindly provided the Fig. **2**.

CONFLICT OF INTEREST

The authors confirm that this chapter content has no conflict of interest.

REFERENCES

[1]   ANDO, H.; ADACHI, M.; UMEDA, K; MATSUURA, A.; NONAKA, M.; UCHIO, R.; TANAKA, H.; MOTOKI, M. Purification and characteristics of a novel transglutaminase derived from microorganisms. Agric Biol Chem 1989; 53(10): 2613-2617.

[2]   NIELSEN, P. M. Reactions and Potential industrial applications of transglutaminase. Review of Literature and Patents. Food Biotechnol 1995; 9(3): 119-156.

[3]   LORAND, L. Transglutaminase - Remembering Heinrich Waelsch. Neurochem Int 2002; 40: 7-12.

[4]   LIN Y., CHAO, M.; LIU, C.; TSENG, M.; CHU, W. Cloning of the gene coding for transglutaminase from *Streptomyces platensis* and its expression in *Streptomyces lividans*. Process Biochem 2006; 41: 519-524.

[5]   MOTOKI, M.; SEGURO, K. Transglutaminase and its use for food processing. Trends Food Sci Technol 1998; 9: 204-210.

[6]   YOKOYAMA, K.; NIO, N.; KIKUCHI, Y. Properties and applications of microbial transglutaminase. Appl Microbiol Biotechnol 2004; 64: 447–454.

[7]   YAN, G., DU, G.; LI, Y; CHEN, J.; ZHONG, J. Enhancement of microbial transglutaminase production by *Streptoverticillium mobaraense:* Application of a two-stage agitation speed control strategy. Process Biochem 2005; 40: 963-968.

[8]   GERBER, U.; JUCKNISCHKE, U.; PUTZIEN, S.; FUCHSBAUER, H. L. A rapid and simple method for the purification of transglutaminase from *Streptoverticillium mobaraense*. Biochem J 1994; 299(3): 825-829.

[9]   JUNQUA, M.; DURAN, R.; GANCET, C.; GOULAS, P. Optimization of microbial transglutaminase production using experimental designs. Appl Microbiol Biotechnol 1997; 48: 730-734.

[10] CUI, L.; DU, G.; ZHANG, D.; LIU, H; CHEN, J. Purification and characterization of transglutaminase from a newly isolated *Streptomyces hygroscopicus*. Food Chem 2007; 105: 612–618.

[11] LIN, S.; HSIEH, Y.; LAI, L.; CHAO, M.; CHU, W. Characterization and large-scale production of recombinant *Streptoverticillium platensis* transglutaminase. J Ind Microbiol Biotechnol 2008; 35: 981–990.

[12] SUZUKI, S.; IZAWA, Y.; KOBAYASHI, K.; ETO, Y.; YAMANAKA, S.; KUBOTA, K.; YOKOZEKI, K. Purification and characterization of novel transglutaminase from *Bacillus subtilis* spores. Biosci Biotechnol Biochem 2000; 64(11): 2344-51.

[13] SOUZA, G. F. V.; FLÔRES, S. H.; AYUB, M. A. Z. Optimization of medium composition for the production of transglutaminase by *Bacillus circulans* BL32 using statistical experimental methods. Process Biochem 2006; 41: 1186-1192.

[14] SOUZA, C. F. V.; VENZKE, J. G.; FLÔRES, S. H.; AYUB, M. A. Z. Enzymatic properties of transglutaminase produced by a new strain of *Bacillus circulans* BL32 and its action over food proteins. LWT - Food Sci Technol 2011; 44: 443-450.

[15] SOARES, L. H. B.; ASSMAN, F.; AYUB, M. A. Z. Production of transglutaminase from *Bacillus circulans* on solid-state and submerged cultivations. Biotechnol Lett 2003; 25: 2029-2033.

[16] RAGKOUSI, K.; SETLOW, P. Transglutaminase-mediated cross-linking of gerq in the coats of *Bacillus subtilis* spores. J Bacteriol 2004; 186: 5567–5575.

[17]   LIN, S.; HSIEH, Y.; LAI, L.; CHAO, M.; CHU, W. Characterization and large-scale production of recombinant *Streptoverticillium platensis* transglutaminase. J Ind Microbiol Biotechnol 2008; 35: 981-990.

[18]   DATE, M.; YOKOYAMA, K.; UMEZAWA, Y.; MATSUI, H.; KIKUCHI, Y. High level expression of *Streptomyces mobaraensis* transglutaminase in *Corynebacterium glutamicum* using a chimeric pro-region from *Streptomyces cinnamoneus* transglutaminase. J Biotechnol 2004; 110: 219-224.

[19]   COLLIGHAN, R.; CORTEZ, J.; GRIFFIN, M. The biotechnological applications of transglutaminases. Minerva Biotecnol 2002; 14: 143–148.

[20]   CORTEZ, J.; ANGHIERI, A.; BONNER, P. L. R.; GRIFFINC, M.; FREDDI, G.Transglutaminase mediated grafting of silk proteins onto wool fabrics leading to improved physical and mechanical properties. Enzyme Microb Technol 2007; 40(7): 1698–1704.

[21]   FONTANA, A.; SPOLAORE, B.; MERO, A.; VERONESE, F. M. Site-specific modification and pegylation of pharmaceutical proteins mediated by transglutaminase. Adv Drug Deliv Rev 2008; 60(1): 13–28.

[22]   ZHU Y.; RINZEMA, A.; TRAMPER, J. Microbial transglutaminase - A Review of its production and application in food processing. J Biol Appl Microbiol Biotecnol 1995, 45: 277-282.

[23]   DONDERO, M.; FIGUEROA, V; MORALES, X.; CUROTTO, E. Transglutaminase effects on gelation capacity of themally induced beef protein gels. Food Chem 2006; 29(4): 546-554.

[24]   TRESPALACIOS, P.; PLA, R. Simultaneous application of transglutaminase and high pressure to improve functional properties of chicken meat gels, Food Chem 2007; 100(1): 264-272.

[25]   ROSSA, P. N.; SÁ, E. M. F.; BURIN, V. M., BORDIGNON-LUIZ, M. T. Optimization of microbial transglutaminase activity in ice cream using response surface methodology. Food Sci Technol 2011; 44: 29-34.

[26]   STEFFOLANI, M. E.; RIBOTTA, P. D.; PEREZ, G. T.; PUPPO, M. C.; LEÓN A. E. Use of enzymes to minimize dough freezing damage. Food Bioprocess Technol 2011; 1-14.

[27]   HUANG, W.; LI, L.; WANGA, F.; WANA, J; MICHAEL, T.; REN, C.; WU, S. Effects of transglutaminase on the rheological and Mixolab thermomechanical characteristics of oat dough. Food Chem 2010; 121: 934–939.

[28]   CUNHA, A. C. Avaliação dos efeitos da aplicação de transglutaminase no processamento de medalhões de salmão. MSc Thesis. 2010, 82p.

[29]   GROSSOWICZ, N.; WAINFAN, E.; BOREK, E.; WAELSCH, H. The enzymatic formation of hydroxamic acids from glutamine and asparagine. J Biol Chem 1950; 187: 111-125.

[30]   FOLK, J. E.; COLE, P. W. Mechanism of action of guinea pig liver transglutaminase. I. Purification and properties of the enzyme identification of a functional cysteine essential for activity. J Biol Chem 1966; 241(23): 5518-5525.

[31]   JEITNER, T. M.; FUCHSBAUER, H.; BLASS, J. P.; COOPER, A. J. L. A sensitive fluorometric assay for tissue transglutaminase. Anal Biochem 2001; 292: 198–206.

[32]   MIWA, N.; YOKOYAMA, K.; WAKABAYASHI, H.; NIO, N. Effect of deamidation by protein-glutaminase on physicochemical. Int Dairy J 2010; 20: 393–399.

[33] HONG, G. P., CHIN, K. B. Effects of microbial transglutaminase and sodium alginate on cold-set gelation of porcine myofibrillar protein with various salt levels. Food Hydrocoll 2010; 24: 444–451.

[34] AHHMED, A. M.; KURODA, R, KAWAHARA, S., OHTA, K.; NAKADE, K.; AOKI, T.; MUGURUMA, M. Dependence of microbial transglutaminase on meat type in myofibrillar proteins cross-linking. Food Chem 2009; 112: 354–361.

[35] KULIK, C.; HEINE, E.; WEICHOLD, O.; MÖLLER, M. Synthetic substrates as amine donors and acceptors in microbial transglutaminase-catalysed reactions. J Mol Catal B-Enzym 2009; 57: 237–241.

[36] TEATHER, R. M.; WOOD, P. J. Use of Congo Red-polysaccharide interactions in enumeration and characterization of cellulolytic bacteria from the bovine rumen. Appl Environm Microbiol 1982; 43(4): 777-789.

[37] KOUKER, G.; JAEGER, K. -E. Specific and sensitive plate assay for bacterial lipases. Appl Environ Microb 1987; 53(1): 211-213.

[38] YOKOYAMA, K.; UTSUMI, H.; NAKAMURA, T., OGAYA, D., SHIMBA, N, SUZUKI, E, *et al.* Screening for improved activity of a transglutaminase from Streptomyces mobaraensis created by a novel rational mutagenesis and random mutagenesis. Appl Microbiol Biotechnol 2010; 87:2087–2096.

[39] SOARES, L. H. B.; ASSMANN, F.; AYUB, MAZ. Purification and properties of a transglutaminase produced by a *Bacillus circulans* strain isolated from the amazon environment. Biotechnol Appl Biochem 2003; 37(3): 295-299.

[40] WU, Y.; TSAI, Y. A rapid transglutaminase assay for high-throughput screening applications. J Biomol Screen 2006; 11: 836-843.

[41] SOKULLUE, B. A. S.; D., BOYACI, I. H.; ÖNER, Z.; KARAHAN A.G.; ÇAKIR, I. *ET AL*. Determination of transglutaminase activity using fluorescence spectrophotometer, Food Biotechnol 2008; 22: 297–310.

[42] PASTERNACK, R.; LAURENT, H.; RÜTH, T.; KAISER, A.; SCHÖN, N., FUCHSBAUER, H. A. fluorescent substrate of transglutaminase for detection and characterization of glutamine acceptor compounds. Anal Biochem 1997; 249: 54–60.

[43] JEITNER, T. M.; DELIKATNYE, E. J. AHLQVIST, J., CAPPERF 18.02; COOPER A. J. L. Mechanism for the inhibition of transglutaminase 2 by cystamine. Biochem Pharmacol 2005; 69(6): 961-970.

Send Orders for Reprints to reprints@benthamscience.net

CHAPTER 10

Keratinases: Detection Methods

Alane Beatriz Vermelho[1,2,*], Ana Maria Mazotto[1] and Sabrina Martins Lage Cedrola[1]

[1]Department of General Microbiology, Institute of Microbiology Paulo de Góes and [2]Biotechnology Center-Bioinovar, Federal University of Rio de Janeiro, Rio de Janeiro, Brazil

Abstract: Keratinases are an important group of enzymes with many industrial applications, including the processing of feather residues, feather meal production, detergent, leather and in the pharmaceutical industry. In this chapter the main methods used to detect their activity are described and discussed. Methods to detect and quantify keratinase activity include spectrophotometric analyses with multiple substrates, zymography and SDS-PAGE containing protein substrates. Focus is on the current status of these biochemical analytical methods.

Keywords: Keratinase, peptidase, keratin, sulfite, serine peptidase, metallo-peptidase, keratin azure.

INTRODUCTION

Microbial keratinases are largely serine or metallopeptidases (EC 3.4.21/24/99.1) produced by various bacteria and fungi [1, 2]. Keratinases degrade keratin which is an insoluble protein cross-linked disulfide, hydrogen and hydrophobic bonds. Keratin is a protective structure found in epidermic appendages such as wool, hair (α-keratin) and feathers (β-keratin) [3, 4]. Currently these enzymes have multiple applications in the pharmaceutical, cosmetic, food, animal feed, leather and 8]. In the literature the use of non specific substrates such as gelatin, casein or azocasein are commonly used to detect keratinase activity.

*Address correspondence to Alane Beatriz Vermelho: Institute of Microbiology Paulo de Góes, Rio de Janeiro, Brazil; Tel: (021)25626743; Fax: (021) 25608344; E-mail: abvermelho@micro.ufrj.br

detergent industries [5, 6]. Besides this, keratinases have been studied due to their ability to inactivate prion and to act as a pesticide against root-knot nematodes [7, 8]. In the literature the use of non specific substrates such as gelatin, casein or azocasein are commonly used to detect keratinase activity.

Methods to detect and quantify keratinase activity include: spectrophometric analyses using azo-keratin [9], keratin azure [10], guinea pig hair [11], feathers and other keratin structures [12, 13].

Electrophoresis of peptidases under non-denaturing conditions in acrylamide gels containing the appropriate proteins entrapped in the gel has also been explored as a general technique to detect keratinase in biological fluids, in cell supernatants and in microorganisms. Heussen and Dowdle [14] introduced the use of SDS–PAGE with gelatin for the detection and identification of plasmin and plasminogen activators. The SDS–PAGE method is based on copolymerization with the protein substrate, which is then degraded by the enzyme in a reaction buffer after separation by electrophoresis. Gelatin, hemoglobin, casein and other proteins have been successfully incorporated into SDS–PAGE gels for qualitative analyses [15-17]. In the literature there are reports describing the use of keratin substrate for keratinase detection [18-20].

Hibino [18] purified a human epidermal keratin and copolymerized it in zymograms, Bernal *et al.* [19] used feather keratin to detect a 240 kDa keratinase by zymography and Kojima *et al.* [20] confirmed that the 27 kDa enzyme purified from *Bacillus pseudofirmus* FA30-01supernatant was a keratinase using SDS-PAGE containing feather meal. Our group has demonstrated the detection of keratinases from different microorganisms using zymography containing feather keratin obtained by DMSO extraction [4, 21-24].

In this chapter we describe the main qualitative and quantitative methods for detecting and measuring keratinases. Additionally, we will also cover methods for detection of sulfite, thiols and disulfide reductase activity that could be involved in the hydrolysis of keratin.

Keratinolytic Activities and Related Activities: Quantitative Methods

Keratinolytic Activity Using Keratin Azure

Principle

Keratin azure, an Azure dye-impregnated sheep's wool keratin, is a commercial substrate for the detection and measurment of keratinase activity. The method is most commonly used in spectrophotometer at a quantitative level but it has been used in culture medium for keratinases detection [10].

Keratinase cleaves Keratin Azure into soluble colored peptides and amino acids according to the scheme below. Keratinase is evaluated through the release of the dye. The color intensity is measured at 595 nm. Thus, a simple visible-wavelength spectrophotometer or a colorimeter could be used.

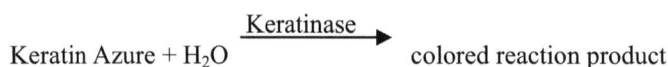

$$\text{Keratin Azure} + H_2O \xrightarrow{\text{Keratinase}} \text{colored reaction product}$$

The enzymatic assay is divided in three parts. The first is the enzymatic reaction itself, the second is the separation of the hydrolytic products from the non-degraded substrate (keratin), and the third is product quantification. Variables that must be defined in the first operation include pH, buffer composition, temperature, and substrate concentration.

Materials

1. Keratinase preparation;

2. Keratin azure (Sigma Prod. No. K-8500);

3. Trizma Base;

4. HCl;

5. Trichloroacetic acid (TCA);

6. Distillated water

7. 10 mL-glass test tubes

8. Pipettes (200 µL and 1000 µL);

9. Glass flask (100 mL);

10. 1mL quartz cuvette

Equipment

1. Spectrophotometer;

2. Analytical balance;

3. Thermostatic bath capable of maintaining temperatures of 30°C and 100°C±0.1 °C;

4. Vortex mixer;

5. Centrifuge;

6. Timer.

Solutions

10mM Tris-HCl pH 8.0

Dissolve 0.121 g of Trizma base in 50 mL of distilled water. Adjust the pH with 1N HCl and complete the volume to 100 mL.

TCA 10%

Dissolve 1 g of trichloroacetic acid in 10 mL of distilled water.

Procedure [25, 26]

1. Prepare six 10-mL glass test tubes for the assay (triplicate sample tubes and triplicate enzyme-substrate blank tubes);

2. Weigh 4 mg of keratin azure into each test tube;

3. Pipette 0.8 mL of 10 mM Tris-HCl pH 8.0 into each test tube;

4. Mix the tubes by swirling and add 0.2 mL of appropriate keratinase preparation to the three sample tubes. The reaction must be done at a low temperature in order to preserve keratinase activity. An ice bath could be used;

5. Add 0.2 mL of distilled water to the three blank tubes. The assay without substrate is used as blank control;

6. Incubate the tubes for 1 hour at 45°C under agitation at 300 rpm;

7. After the reaction period, the reaction is stopped by separation of the hydrolytic products from the non hydrolyzed substrate by centrifugation. Centrifuge all tubes at 15000g for 15 min;

    Obs.: In this step some authors introduced a modification: the addition of 0.5 mL of 10% TCA to the sample tubes in order to precipitate the enzyme and the non degraded substrate. In the blank tubes the addition of 0.5 mL of 10% TCA occurs before the reaction. Then after centrifugation the colored hydrolytic products are measured [27, 28].

Transfer the supernatants to 1 mL cuvettes and measure the absorbance of the supernatant fluid at 595 nm;

Subtracting the mean blank values from the mean sample values gives the amount of peptide azo-dye liberated during hydrolysis;

Determine the activity units, where a unit is defined as that amount of enzyme required to produce an increase of 0.01 absorbance units (595 nm) in an hour.

Calculations

$$\Delta A_{595nm}\, Test = A_{595nm}\, Sample - A_{595nm}\, Blank$$

$$U = \Delta A_{595nm}\, Test/0.01$$

$$U/mg = U/mg\ enzyme$$

$$U/mL = U/enzyme\ volume\ used\ (mL)$$

Example

Bacillus subtilis AMR (isolation was describe by Mazotto *et al.*, 2010) was cultivated in feather medium containing 0.06 M $Na_2HPO_4.7H_2O$ and 0.04 M KH_2PO_4, pH 8.0, 0.5mM $CaCl_2$, 0.5mM $MgSO_4$, 0.5mM $MnCl_2$, 0.05% sucrose and 1.5% chicken feathers. After 6 days of incubation at room temperature, the culture was centrifuged at 300rpm and the supernatant was used for the keratinase assay as describe above (Fig. **1**).

Figure 1: Blank and sample tubes containing Keratin Azure.

Instead of keratin azure other keratinous substrates can be used such as feather [13], keratin extract from feather [21, 29] or azokeratin [27].

Sulfite Determination Using ρ-Rosaniline

Principles

One of the keratin degradation mechanisms includes the disulphide bond cleavage by sulfite. This mechanism, named sulfitolysis was described for the first time for dermatophyte fungi [30].

In this sulfite detection method, the sulfite reacts with ρ-rosaniline hydrochloride and formaldehyde forming a purple-colored sulfonic acid derivative of ρ-rosaniline. The color intensity is proportional to the sulfite concentration (Fig. **2**). The sulfite concentration will be measured spectrophometrically, measuring absorbance at 562 nm.

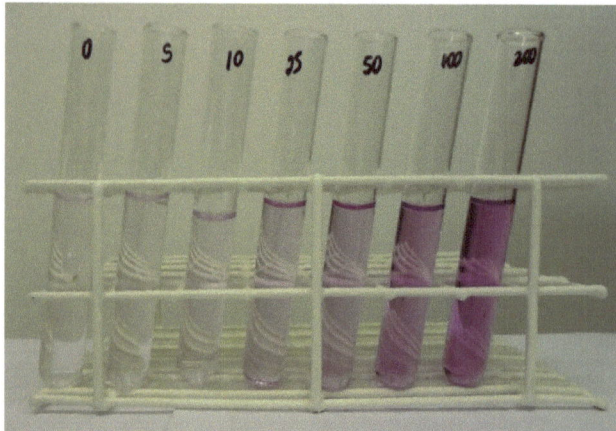

Figure 2: Tubes of calibration curve.

Materials

1.   ρ-rosaniline;

2.   HCl;

3.   Formaldehyde;

4.   Sodium sulfite;

5.   Distilled water;

6.   10 mL-glass test tubes;

7.   Pipettes (100 μL, 1 mL and 5 mL);

8.   Glass flask (100 mL);

9.   5 mL glass cuvette.

Equipment

1. Spectrophotometer;

2. Analytical balance;

3. Thermostatic bath capable of maintaining temperatures of 30°C and 100°C±0.1 °C;

4. Vortex mixer;

5. Timer.

Preparation of Solutions

0.04% ρ-Rosaniline HCl Solution

Weigh 0.04g of ρ-rosaniline hydrochloride.

Add a small volume of distilled water with 3 mL of HCl and mix until dissolved completely.

Add 3 mL of HCl and complete the volume to 100mL.

This solution has a pale yellow color; it is stable against light and can be stored for at least three months.

0.2% Formaldehyde Solution

Dilute 0.2 mL of formaldehyde in 100 mL of distilled water. Store the solution at room temperature.

0.1mg/mL Sodium Sulfite Solution

Dissolve 1 mg of sodium sulfite in 10 mL of distilled water.

Procedure [31]

Calibration Curve

Set up a series of twenty-one 10 mL glass test tubes, enough for triplicate tubes for each dilution of the calibration curve;

Pipette the following reagents into the test tubes.

Tube	Sodium Sulfite Solution (mL)	Distilled Water (mL)	Sodium Sulfite (µg)
1	0	0.5	0
2	0.005	0.495	0.5
3	0.010	0.490	1
4	0.025	0.475	2.5
5	0.050	0.450	5
6	0.100	0.400	10
7	0.200	0.300	20

1. Add 0.5 mL of ρ-rosaniline HCl solution to the tubes;

2. Add 0.5 mL of formaldehyde solution to the tubes;

3. Place the tubes in a vortex mixer for 2 sec.;

4. Pipette 3.5 mL of distilled water into each tube;

5. Leave the tubes at room temperature for 10 min. The solution color changes from light pink to purple with increasing sodium sulfite concentration (Fig. **2**);

6. Transfer 3 mL of the final solution to a 3.5 mL cuvette;

7. Measure absorbance at 562 nm against tube 1 (water instead of sodium sulfite) as a reference, calibrating this first absorbance to zero;

8. Construct a calibration curve by plotting the mean absorbance values observed for each sodium sulfite in the standard sample in µmol (Fig. **3**). Check the coefficient of determination. It should be higher than 0.9900 for accurate determination of sodium sulfite content in samples with unknown content.

Assay for Sample Sulfite Determination

Get three more 10-mL glass test tubes (triplicate sample tubes);

Pipette 0.5 mL of sample (*e.g.,* culture supernatant) into the three sample tubes;

Repeat steps 3-9 above.

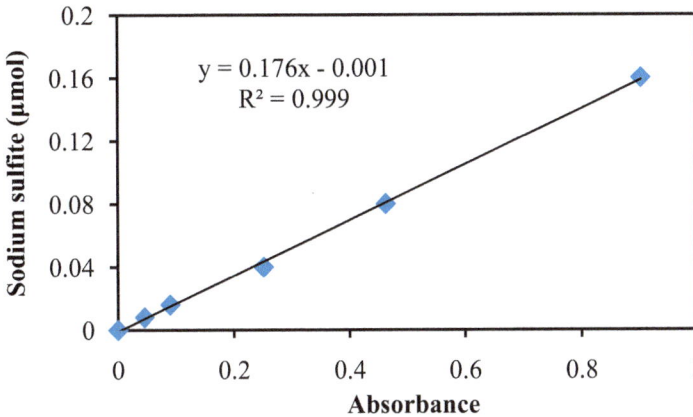

Figure 3: Example of a calibration curve for quantification of sulfite.

Calculation of Sulfite Concentration

After plotting the standard curve of absorbance as a function, use the equation to determine the unknown sulfite concentrations in μmol.

$$\text{Sulfite (μmol)} = \alpha \times abs - \beta$$

Where:

α is the angular coefficient of the calibration curve (in the example, α=0.1769);

abs is the mean absorbance values registered for samples;

β is the linear coefficient (in the example, β = 0.0011).

Example

Bacillus subtilis AMR was cultivated in feather medium (0.06 M $Na_2HPO_4.7H_2O$ and 0.04 M KH_2PO_4, pH 8.0, 0.5mM $CaCl_2$, 0.5mM $MgSO_4$, 0.5mM $MnCl_2$, 0.05% sucrose and 1.5% chicken feathers) for 6 days at room temperature under constant agitation (300rpm). After incubation, the culture was centrifuged and the supernatant was used for sulfite determination as describe above (Fig. **4**).

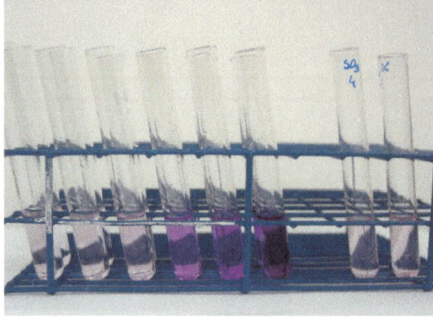

Figure 4: Calibration curve used in the test (left) and the sample tubes (right).

Disulfide Reductase Activity

Principle

The cleavage of the cystine bonds may have a significant influence on keratin degradation. Besides sulfitolysis, Yamamura *et al.* [32] described another keratin degradation mechanism with a synergistic action of two enzymes (peptidases and a disulfide reductase) in a cooperative system which could degrade keratin [32, 33, 34]. The test described below must be used only for purified enzymes.

The enzyme disulfide reductase mediates the reduction of disulfide bonds [35]. They have an active site containing two cysteines, arranged in a C-X-X-C motif, which are either in the reduced state forming two thiols or in the oxidized state forming an intramolecular disulfide bond. This motif represents the active site of enzymes that interact directly with cysteines or cystines in the target molecules (Fig. **5**).

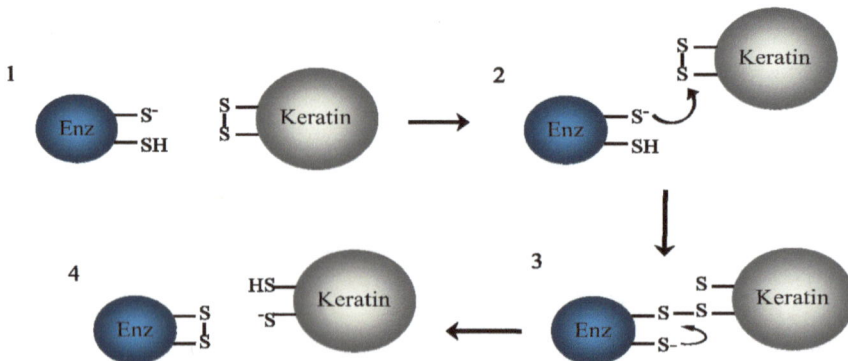

Figure 5: Mechanisms of cleavage of a disulfide bond by a disulfide reductase (Enz) (adapted from Maeda *et al.* [36] and Fabianek *et al.* [37]). (1) Active site of disulfide reductase before reaction. The cysteines of the substrate are in the oxidized state forming disulfide bonds. (2) The N-terminal cysteine makes a nucleophilic attack on the keratin disulfide bond. (3) An

intermolecular mixed disulfide intermediate is formed, which is subjected to intramolecular attack from the C-terminal cysteine (4) to release the reduced keratin and oxidized disulfide reductase.

In this method the enzyme disulfide reductase catalyzes the reduction of oxidized glutathione (glutathione disulfide, GSSG) to glutathione (GSH) exposing two sulfhydryl groups (Fig. **6**). The content of –SH produced can be assayed using Ellman's method [38], in which 5,50-dithio-2-nitrobenzoate (DTNB) is used to react with the SH groups to produce a yellow substance with a maximum absorbance at 412 nm [39].

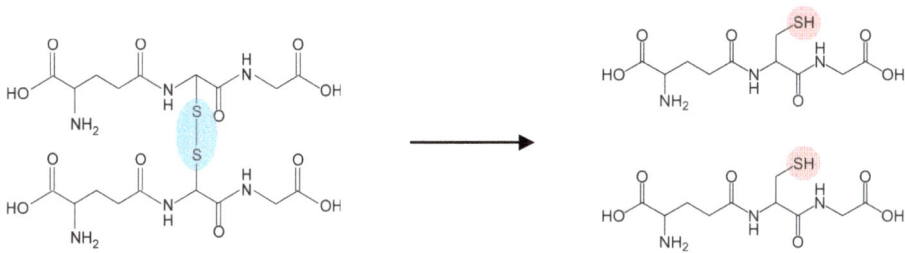

Figure 6: Reduction of disulfide glutathione.

Materials

1. Enzyme preparation;

2. DTNB (5,50-dithio-2-nitrobenzoate);

3. Oxidized glutathione;

4. Na_2HPO_4;

5. KH_2PO_4;

6. PMSF (phenylmethanesulfonylfluoride);

7. EDTA;

8. Distillated water.

9. Glass test tubes;

10. Plastic microtubes (1.5 mL, polypropylene);

11. Pipettes (10 µL, 100 µL and 1 mL);

12. Glass flask (100 mL);

13. Glass cuvette (1 mL);

Equipments

1. Spectrophotometer;

2. Analytical balance;

3. Thermostatic bath capable of maintaining temperatures of 30°C and 100°C±0.1 °C;

4. Vortex mixer;

5. Chronometer.

Solutions

Phosphate Buffer pH 8.0

9.5 mL of Na_2HPO_4 solution + 0.5 mL KH_2PO_4 solution;

Na_2HPO_4 solution -9.474 g/l;

KH_2PO_4 solution -9.078 g/l.

2 mM Oxidized Glutathione

Dissolve 12.25 mg of oxidized glutathione in 10mL of phosphate buffer pH 8.0.

0.2 M PMSF

Dissolve 1.742 g of PMSF in 50mL of ethanol

12 mg/mL DTNB

Dissolve 12 mg DTNB in 1 mL of phosphate buffer pH 8.0 containing 1mM EDTA.

Procedure [39, 40]

1. Prepare a series of six 1.50-mL polypropylene test microtubes for an enzyme preparation to be analyzed (enough for triplicate sample tubes and triplicate enzyme-substrate blank tubes);

2. Add 250 μL of phosphate buffer pH 8.0 to the each of the six tubes;

3. Add 250 μL of appropriate enzyme preparation (preferentially dialyzed or purified sample) to the tubes;

4. Pipette 15 μL of PMSF into all the tubes;

5. Add 500 μL of oxidized glutathione to the three sample tubes. Replaced the substrate solution for phosphate buffer pH 8.0 in the three blank tubes;

6. Place the tubes in a vortex mixer;

7. Incubate the tubes at 37°C for 30 min.;

8. Centrifuge the tubes at 1000g for 5 min;

9. Transfer 750 μL of reaction mixture supernatant to other microtubes;

10. Add 25 μL of DTNB;

11. Place the tubes in a vortex mixer for 2 seg.;

12. Leave the tubes at room temperature for 5 min to stabilize the color development;

13. Measure the reaction mixture spectrophotometrically at 412 nm by measuring the yellow-colored sulfide formed upon reduction of DTNB;

14. Subtracting the mean blank value from the mean sample value to obtain the amount of -SH liberate during hydrolysis;

15. Calculate the amount of sulfhydryls in the sample from the molar extinction coefficient of TNB ($14,150 \text{ M}^{-1}\text{cm}^{-1}$);

16. Determine the activity units, where a unit of disulfide bond-reducing activity was defined as the amount of enzyme that catalyzes the formation of 1 μmole of sulfhydryl per min;

Calculations

$$\Delta A_{412nm} \text{ Test} = A_{412nm} \text{ Sample} - A_{412m} \text{ Blank}$$

Calculate the sulfhydryl concentration in moles/liter (=M) using the following formula:

$$c = \frac{A}{bE}$$

Where:

A = absorbance (ΔA_{412nm} Test);

b = path length in centimeters (1 cm);

E = molar absorptivity of TNB (in this buffer system at 412 nm is 14,150).

For the calculation of activity, use the equation below.

$$\mathbf{U = c * 10^6 * 30}$$

$$\text{U/mL} = \text{U} * 1000/250$$

Where:

c = sulfhydryl concentration in moles/liter;

10^6 is the conversion to μmol;

30 corresponds to the incubation time.

Example

Supernatant of *Bacillus subtilis* AMR cultivated in feather medium (0.06 M $Na_2HPO_4.7H_2O$ and 0.04 M KH_2PO_4, pH 8.0, 0.5mM $CaCl_2$,

0.5mM $MgSO_4$, 0.5mM $MnCl_2$, 0.05% sucrose and 1.5% chicken feathers) for 6 days at room temperature under constant agitation (300rpm) was used in this assay. After incubation, the culture was centrifuged and the supernatant was used for determination of disulfide reductase activity as describe above. The mean blank value was 0.004 and the mean sample value was 0.012.

$$\Delta A_{412nm} \, Test = 0.012 - 0.004 = 0.008$$

$$C = 0.008/1 * 14,150 = 5.65 * 10^{-7}$$

$$U = 5.65*10^{-7} * 106 * 30 = 16.96$$

$$U/mL = 16.96 * 1000/250 = 67.8 \ U/mL$$

Keratinolytic Activities: Qualitative Methods

Enzymography

Principle of the Method

Zymography is used to evaluate the ability of an enzyme to degrade a specific substrate using sodium dodecyl sulfate polyacrylamide gel electrophoresis (SDS-PAGE). The keratinase preparation is incubated with the substrate for variable intervals of time and after incubation the reaction mixture is applied in SDS-PAGE. SDS-PAGE causes the separation of macromolecules in an electric field and thus it is used for separating proteins LaemmLi [41]. SDS (also called lauryl sulfate) is an anionic detergent that confers to the proteins a negative charge within a wide pH range. The amount of SDS attached to the protein is in proportional to its molecular mass. This procedure allows the protein migration based only on its molecular mass and not due to the charge in an electric field. If the keratinase cleave the substrate, novel protein fragments will appear or the intensity of the keratin substrate in the mixture reaction will decrease.

Materials

1. Acrylamide;

2. Bisacrylamide;

3.   Trizma Base;

4.   Keratin solution;

5.   Glycerol;

6.   Ammonium persulfate;

7.   TEMED;

8.   Distilled water;

9.   SDS;

10. Bromophenol blue;

11. Sample to degrade keratin (culture supernatants);

12. Molecular mass standards (SDS-PAGE standards);

13. Triton X-100;

14. Coomassie blue;

15. Methanol;

16. Acetic acid;

17. Electrophoresis apparatus and comb;

18. Gel-loading pipette tips or Hamilton syringe;

19. Pipettes and flasks;

20. Constant voltage power supply;

21. Sealed plastic container.

Reagents and Solutions

Acrylamide / Bisacrylamide Solution

30% acrylamide;

0.8% bisacrylamide;

Dissolve the reagents in distilled water and filter twice. Store up to 6 months at 4°C.

Segregation Gel Buffer (1.5M Tris pH 8.0)

Dissolve 18.15 g trizma base in ~90 mL distilled water. Adjust pH to 8.8 with 6N HCl and adjust volume to 100 mL with distilled water. Store up to 1 month at 4°C.

Keratin Solution

Chicken feathers obtained from poultry waste were washed extensively with water and detergent, dried at 60°C overnight, delipidated with chloroform: methanol (1:1, v/v) and dried again at 60°C. The DMSO extraction method Wawrzkiewicz *et al.* [21] produces a feather keratin suspension that when dried gives a keratin powder. Briefly, 10 g of feathers were heated with a reflux condenser at 100°C for 80–120 min with 500 mL of DMSO. Keratin was then precipitated by the addition of two volumes of acetone and maintained at 4°C for 24–48 h. The keratin precipitates were collected by centrifugation (2,000g/15 min), washed twice with distilled water and dried at 4°C. A white powder was obtained for use as a keratin standard in feather degradation studies [4, 6].

10% (w/v) Ammonium Persulfate

Dissolve 1 g ammonium persulfate in a final volume of 10 mL water. Store up to 1 month at -20°C.

10% SDS

Dissolve 10 g SDS in 100 mL distilled water. Store up to 3 months at room temperature.

Stacking Gel Buffer (0.5M Tris-HCl pH 6.8)

Dissolve 6 g trizma base in ~90 mL distilled water. Adjust pH to 6.8 with 6 N HCl and adjust volume to 100 mL with distilled water. Store up to 1 month at 4°C.

Running Buffer (4x)

Dissolve 3 g Tris base, 14 g glycine and 1 g SDS in ~900 mL distilled water. Adjust volume to 1000 mL with distilled water. Store up to 1 month at 4°C. Dilute to 4x with distilled water as required.

Sample Loading Buffer

Prepare: Add 8 mL SDS 10%, 8 mL glycerol 50%, 5 mL 0.5M Tris pH 6.8 to 16 mL distilled water and add 2 mL mercaptoethanol and some bromophenol blue. Store up to 1 month at −20°C.

Final solution: 125mM Tris, pH 6.8, 4% SDS, 20% glycerol, 0.002% bromophenol blue supplemented with 5% (v/v) b-mercaptoethanol (Mazotto *et al.*, 2011).

Enzyme Renaturing Solution (10% Triton X-100)

Add gradually 100 mL Triton X-100 to 700 mL distilled water under gentle agitation. Adjust volume to 1000 mL with water. Store up to 3 months at 4°C. Dilute to 4x with distilled water as required.

Staining Solution (Coomassie Blue)

Stock solution (Dissolve 0.5 g Coomassie brilliant blue R-250 in 25 mL distilled water).

Mixt 12.5 mL stock solution with 50 mL methanol and 10 mL acetic acid, and adjust volume to 100 mL with distilled water. Store up to several months at room temperature.

Destaining Solution 1

Mix 500 mL methanol with 100 mL acetic acid and adjust volume to 1000 mL with distilled water. Store up to several months at room temperature.

Destaining Solution 2

Combine 50 mL methanol with 70 mL acetic acid and adjust volume to 1000 mL with distilled water. Store up to several months at room temperature.

Procedures

1. The percentages of acrylamide gel are in Table **1**.

    The choice of acrylamide percentage depends on the molecular weight of the peptidases to be visualized. Proteins with higher molecular masses resolve better on lower percentage gels, while those with lower molecular masses resolve better on higher percentage gels.

2. Prepare the gel by mixing the following:

Table 1: Segregation gel

Solution / Reagents	7.5%	10%	12.5%	15%
Segregation gel buffer (pH8.8)	3 mL	3 mL	3 mL	3 mL
Acrylamide / bisacrylamide solution	3 mL	4 mL	5 mL	6 mL
Glycerol	1.2 mL	1.2 mL	1.2 mL	1.2 mL
10% (w/v) ammonium persulfate (APS)*	120 μL	120 μL	120 μL	120 μL
TEMED*	20 μL	20 μL	20 μL	20 μL
Distilled water	4.8 mL	3.8 mL	2.8 mL	1.8 mL
10%SDS	120 μL	120 μL	120 μL	120 μL

1. Add consecutively 120 μL of APS and 20 μL of TEMED to the gel solution. The gel will polymerize fairly quickly, so do not add these until you are sure you are ready to pour them into the electrophoresis apparatus. Put the segregation gel between the electrophoresis plates at a height 3 cm from the top, cover with distilled water and then wait for the polymerization of the acrylamide and bisacrylamide.

2. Prepare a stacking gel by mixing the following:

Table 2: Stacking gel

Acrylamide / bisacrylamide solution	830 μL
Stacking gel buffer	1.25 mL
Distilled water	2.82 mL
10% SDS	50 μL
10% (w/v) ammonium persulfate	60 μL
TEMED	20 μL

Discard the water and immediately pour the stacking gel solution on top of the segregation gel. Insert the well comb, taking care not to trap bubbles under the comb. Allow the stacking gel to polymerize 15 to 20 min at room temperature.

3.   Load samples (blend with sample loading buffer and samples) and molecular mass standards into appropriate wells using gel-loading pipette tips or a Hamilton syringe.

4.   Electrophorese using a constant voltage power supply at 170 V until the bromophenol blue reaches the bottom of the gel. A 12.5% acrylamide gel should take ~120 min to complete.

5.   After the electrophoresis run, disassemble the gel apparatus and add the staining solution, leave for about 1 h at room temperature. Discard the staining solution and add destaining solution 1 for 1 h or until clear bands are visible against the dark blue background. Discard this solution and add destaining solution 2.

The gel can be stored in destaining solution 2 for up to a few months at room temperature.

6.   Compare the height of the molecular mass standard bands with the samples applied.

Example

The *Bacillus subtillis* SLC was cultivated on phosphate buffered medium (0.06 M $Na_2HPO_4.7H_2O$ and 0.04 M KH_2PO_4, pH 8.0) supplemented with 1% native feather as the only nitrogen and carbon source for 5 days at room temperature on a rotary shaker (300 rpm). After incubation, the media were centrifuged at 2,000g for 20 min at 4°C and the supernatant solutions were concentrated 25-fold by dialysis (cut-off 9 kDa) against PEG 4000 overnight at 4°C. The concentrated supernatant was used as enzyme extract in the enzymography. 20 μL enzyme extract (30 μg mL^{-1}) was mixed with an equal volume of keratin solution (0.25 mg/mL). These reaction mixtures were incubated for 1 h at 37 °C. Reaction mixtures (20 μL) were added to 20 μL sample loading buffer and boiled at 100 °C for 5 min. Protein degradation (keratin hydrolysis) was then analyzed on 15% SDS-PAGE [41]. Electrophoresis was carried out at 170 V for 2 h, and then the gel was stained with staining solution (about 2 h) and destaining solution 1 was added. After the appearance of protein bands, the gel was kept in destaining solution 2 [42]. As control for possible degradation of the proteinaceous substrates independent of proteolytic enzymes, aliquots of the concentrated supernatant were heat-inactivated with the substrates before incubation. In addition, a second control for keratin substrate was carried out by replacing concentrated supernatants with the same volume of buffer [17].

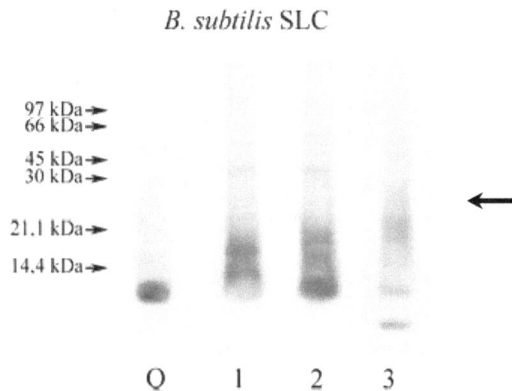

Figure 7: Feather keratin degradation by extracellular keratinases of *B. subtilis* SLC. Feather keratin degradation profiles were analyzed by 15% SDS-PAGE, and gels were stained with coomassie blue. Gel Strip Q, feather keratin substrate; Strip 1, Crude enzyme extract (culture supernatant), Strip 2, enzyme extract + feather keratin in Time 0 (control); and Strip 3, the same solution as Strip 2 but incubated for 1 h at 37°C.

After 1 h of incubation of feather keratin with supernatant solutions from a sample of the *B. subtilis* SLC, a 10 kDa band characteristic of feather keratin was almost totally degraded and a band with lower molecular weight was found, indicating that the keratin had been broken down (Fig. **7**).

Zymography

Principle of the Method

Zymography is a simple, sensitive, and quantifiable technique that is widely used to detect peptidases [14, 43, 44] and allows visualization of the number and approximate molecular masses of peptidases in a sample on the basis of hydrolysis of a protein substrate within the gel. It is a two-stage technique involving protein separation by electrophoresis, followed by detection of enzyme activity in polyacrylamide gels under non-reducing (no treatment of reducing agents DTT or b-mercaptoethanol) conditions. The technique is particularly useful for analyzing peptidase compositions of complex biological samples because visualization depends directly on proteolytic activity. These enzymes should have the ability to renature after removal of SDS and to exert proteolytic activity in the co-polymerized substrate [45, 46]. The standard method for zymography is based on the use of SDS-polyacrylamide gels co-polymerized with a protein substrate, in particular gelatin [47], casein [43, 48], fibrin [44] or keratin [4, 49].

Peptidase activity in zymograms is visualized as a clear band of proteolysis that appears against the dark-blue background (after Coomassie Brilliant Blue staining) of undigested protein substrate [44]. On the basis of molecular weight markers, the molecular weight of the proteolytic band can be determined, and by comparison with recombinant proteins and the use of specific peptidase inhibitors the type of peptidase can be established.

To provide a uniformLy dispersed substrate, the peptide (keratin) is included in the standard polyacrylamide gel solution and the gel is then poured and polymerized as usual. Since the substrate is entrapped in the pores of the polyacrylamide gel, it does not migrate out of the gel during electrophoresis. Peptidase samples are denatured with sodium dodecyl sulfate (SDS), but not boiled or reduced, and then electrophoresed. The separated peptidases are renatured within the gel by replacing the SDS with a nonionic detergent such as Triton X-100. The gel is then incubated in a buffer suitable for enzymatic activity, allowing the renatured peptidase to digest the gel-bound substrate protein in a zone around their electrophoresed position. These zones of digestion are visualized using Coomassie brilliant blue R250, with the areas of digestion appearing clear [50].

Materials and Equipments

1. Acrylamide;

2. Bisacrylamide;

3. Trizma Base;

4. Keratin;

5. Glycerol;

6. Ammonium persulfate;

7. TEMED;

8. Distilled water;

9. SDS;

10. Bromophenol blue;

11. Sample to degrade keratin (culture supernatants);

12. Molecular mass standards (SDS-PAGE standards);

13. Triton X-100;

14. Coomassie blue;

15. Methanol;

16. Acetic acid;

17. Electrophoresis apparatus and comb;

18. Gel-loading pipette tips or Hamilton syringe;

19. Pipettes and flasks;

20. Constant voltage power supply;

21. Sealed plastic container.

Reagents and Solutions

Acrylamide / Bisacrylamide Solution

30% acrylamide;

0.8% bisacrylamide;

Dissolve the reagents in distilled water and filter twice. Store up to 6 months at 4°C.

Separation Gel Buffer (1.5M Tris pH 8.0)

Dissolve 18.15 g trizma base in ~90 mL distilled water. Adjust pH to 8.8 with 6N HCl and adjust volume to 100 mL with distilled water. Store up to 1 month at 4°C.

Keratin Substrate

Chicken feathers obtained from poultry waste were washed extensively with water and detergent, dried at 60°C overnight, delipidated with chloroform: methanol (1:1, v/v) and dried again at 60°C. The Wawrzkiewicz *et al.* [21] method was modified to produce keratin powder from the lipid free dried feathers. Briefly, 10 g of feathers were heated with a reflux condenser at 100°C for 80–120 min with 500 mL of DMSO. Keratin was then precipitated by the addition of two volumes of acetone and maintained at 4°C for 24–48 h. The keratin precipitates were collected by centrifugation (2,000 g/15 min), washed twice with distilled water and dried at 4°C. A white powder was obtained for qualitative and quantitative biochemical analyses related to keratinases activity and as a keratin standard in feather degradation studies [4, 6]. The keratin powder (0.07 g) is added to 12.23 mL of polyacrylamide gel.

10% (w/v) Ammonium Persulfate

Dissolve 1 g ammonium persulfate in a final volume of 10 mL water. Store up to 1 month at -20°C.

10% SDS

Dissolve 10 g SDS in 100 mL distilled water. Store up to 3 months at room temperature.

Stacking Gel Buffer (0.5M Tris-HCl pH 6.8)

Dissolve 6 g trizma base in ~90 mL distilled water. Adjust pH to 6.8 with 6 N HCl and adjust volume to 100 mL with distilled water. Store up to 1 month at 4°C.

Running Buffer (4x)

Dissolve 3 g Tris base, 14 g glycine and 1 g SDS in ~900 mL distilled water. Adjust volume to 1000 mL with distilled water. Store up to 1 month at 4°C. Dilute to 4x with distilled water as required.

Sample Loading Buffer

Dissolve 1.2 g SDS and 0.1 g bromophenol blue in 2.5 mL stacking gel buffer and add 7.2 mL glycerol. Adjust volume to 15 mL with water. Store up to 1 month at −20°C.

Enzyme Renaturing Solution (10% Triton X-100)

Add gradually 100 mL Triton X-100 to 700 mL distilled water under gentle agitation. Adjust volume to 1000 mL with water. Store up to 3 months at 4°C. Dilute to 4x with distilled water as required.

Staining Solution (Coomassie Blue)

Stock solution (Dissolve 0.5 g Coomassie brilliant blue R-250 in 25 mL distilled water).

Mix 12.5 mL stock solution with 50 mL methanol and 10 mL acetic acid, and adjust volume to 100 mL with distilled water. Store up to several months at room temperature.

Destaining Solution 1

Mix 500 mL methanol with 100 mL acetic acid and adjust volume to 1000 mL with distilled water. Store up to several months at room temperature.

Destaining Solution 2

Combine 50 mL methanol with 70 mL acetic acid and adjust volume to 1000 mL with distilled water. Store up to several months at room temperature.

Procedures

1. Choose an appropriate concentration of Acrylamide / bisacrylamide and prepare according to Table **1**. The choice of acrylamide percentage depends on the molecular mass of the peptidase.

2. Prepare the gel by mixing the following:

Table 1: Running gel

Percentage of Gel Solution / Reagents	7.5%	10%	12.5%	15%
Separation gel buffer (pH8.8)	3 mL	3 mL	3 mL	3 mL
Acrylamide / bisacrylamide solution	3 mL	4 mL	5 mL	6mL
Glycerol	1.2 mL	1.2 mL	1.2 mL	1.2 mL
10% (w/v) ammonium persulfate (APS)*	120 μL	120 μL	120 μL	120 μL
TEMED	20 μL	20 μL	20 μL	20 μL
Keratin substrate	0.07 g	0.07 g	0.07 g	0.07 g
Distilled water	4.8 mL	3.8 mL	2.8 mL	1.8 mL
10%SDS	120 μL	120 μL	120 μL	120 μL

3. Stir the solution (without APS and TEMED) continuously for 1 hour. Then centrifuge at 2000g for 15 min to remove the nondissolved substrate.

4. Add consecutively 120 µL of APS and 20 µL of TEMED to the gel solution. The gel will polymerize fairly quickly. Put the segregation gel between electrophoresis plates at a height of 3 cm from the top, cover with distilled water and then wait for the polymerization of the acrylamide and bisacrylamide.

Prepare a stacking gel by mixing the following:

Table 2: Stacking gel

Acrylamide / bisacrylamide solution	830 µL
Stacking gel buffer	1.25 mL
Distilled water	2.82 mL
10% SDS	50 µL
10% (w/v) ammonium persulfate	60 µL
TEMED	20 µL

Discard the water and immediately pour the stacking gel solution on top of the segregation gel. Insert the well comb, taking care not to trap bubbles under the comb. Allow the stacking gel to polymerize for 15 to 20 min at room temperature.

5. Load samples (blend with sample loading buffer and culture supernatants) and molecular mass standards into appropriate wells using gel-loading pipette tips or a Hamilton syringe.

6. Run gels at 170V (constant voltage) until the bromophenol blue dye reaches the bottom of the gel. A 12.5% acrylamide gel should take ~120 min to complete.

7. After the electrophoresis run, disassemble the gel apparatus and cut the slot where the standard of molecular weight was applied and go to the staining step (10). Place the remainder of the gel in a container filled with 25 mL enzyme renaturing solution. Wash 15 min at room temperature with smooth agitation on a shaker. Repeat this procedure twice using a fresh solution each time. This procedure removes the SDS of the gel.

8. Then discard the enzyme renaturing solution and add 30 mL of incubation buffer and incubate overnight at 37°C.

9. The incubation time and buffer will depend on the sample of enzyme used. The container should be airtight to prevent evaporation.

    Discard buffer incubation and add the staining solution and leave for about 1 h at room temperature. Discard the staining solution and add destaining solution 1 and leave for 1 h or until clear bands are visible against the dark blue background. Discard this solution and add destaining solution 2.

    The gel can be stored in destaining solution 2 for up to a few months at room temperature.

10. Compare the height of the bands with the molecular mass standards with the area of digestion of the sample to quantify the mass of enzymes.

Example

The *Bacillus subtillis* SLC was cultivated on phosphate buffered medium (0.06 M $Na_2HPO_4.7H_2O$ and 0.04 M KH_2PO_4, pH 8.0) supplemented with 1% native feather as the only nitrogen and carbon source for 5 days at room temperature on a rotary shaker (300 rpm). After incubation, the media were centrifuged at 2,000g for 20 min at 4°C and the supernatant solutions were concentrated 25-fold by dialysis (cut-off 9 kDa) against PEG 4000 overnight at 4°C. The concentrated supernatant was used as the enzyme extract to analyze keratinase activity. Enzyme extract (40 µL) and sample loading buffer (60 µL) were mixed and 10 µL was applied in the gel (12.5%) containing keratin. As described above.

The ability of the extracellular peptidases to degrade the keratin substrate was evaluated using keratin co-polymerized with 12.5% sodium dodecyl sulfate–polyacrylamide gels (SDS–PAGE). Fig. **8** shows that *B. subtilis* SLC presented bands with an apparent molecular mass of 80, 54, 30 and 15 kDa.

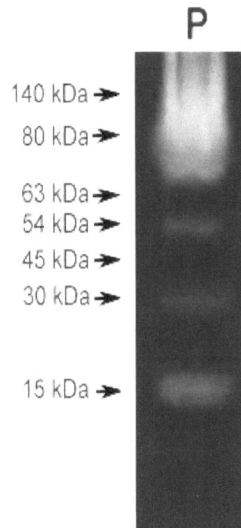

Figure 8: Zymograms with co-polymerized feather keratin. Enzyme extracts of *B. subtilis* SLC were prepared as described above. Gel strips were incubated for 48 h at 37 °C in 0.5 M Tris-HCl, pH 7.4. The molecular masses of the peptidases, expressed in kDa, are shown on the left.

Detection of Extracellular Keratinase on Agar Plates

Principle of the Method

For the isolation and screening of new microbes for keratinase production a rapid, efficient and sensitive technique for the enzyme detection is required. In this context the keratin-agar plate detection method could be a useful, easy and qualitative method [51].

The extracellular keratinases obtained from agar nutrient were added to the wells of the keratin agar. Degradation halos could be visualized around the wells.

Procedure

Nutrient Medium

Composition: 2% (w/v) KCl, 2% (w/v) sucrose, 0.5% (w/v) peptone, 0.5% (w/v) yeast extract and 2% (w/v) agar in distilled water. This solution was autoclaved, cooled to approximately 50°C and then 20 mL were poured onto each plate and allowed to harden.

Keratin-Agar

Composition: 0.01% (w/v) yeast extract, 2% (w/v) agar and 2% (w/v) keratin substrate in distilled water. This solution was autoclaved, cooled to approximately 50°C and then 20 mL of the suspension was poured gently into a Petri dish and left to harden. A circular well is made n the agar plates with a sterile Durhan tube. The enzyme extract containing the keratinases is put into the well (s).

Staining Solution (Coomassie Blue)

Stock solution (Dissolve 0.5 g Coomassie brilliant blue R-250 in 25 mL distilled water).

Mix 12.5 mL of stock solution to 50 mL methanol and 10 mL acetic acid, and adjust volume to 100 mL with distilled water. Store up to several months at room temperature.

Destaining Solution 1

Mix 500 mL methanol, 100 mL acetic acid and adjust volume to 1000 mL with distilled water. Store up to several months at room temperature.

Destaining Solution 2

Combine 50 mL methanol with 70 mL acetic acid and adjust volume to 1000 mL with distilled water. Store up to several months at room temperature.

Example

The microorganism used was a *Bacillus subtilis* [49]. The *Bacillus* sp was grown in nutrient medium for 48h at room temperature. Then a loopful of the culture growth was placed in 0.85% (w/v) saline and 500 μL of this mixture was placed in the wells of agar keratin. After inoculation, the plate was incubated at 37°C and observed daily for five days. Extracellular keratinases detection was done after staining with staining solution (coomassie blue) although destaining may require the use of destaining solution 1. Degradation halos were formed around the well (Fig. **9**).

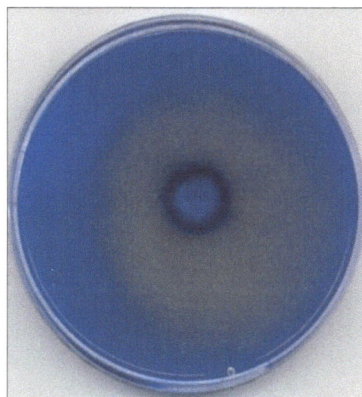

Figure 9: Keratinolytic activity on agar plates containing keratin as substrate.

ACKNOWLEDGEMENTS

The authors wish to acknowledge the Brazilian National Council for Scientific and Technological Development (CNPq), Carlos Chagas Filho Foundation for Research Support in the State of Rio de Janeiro (FAPERJ), Coordination for the Improvement of Higher Education Personnel (CAPES). The authors are grateful to Denise da Rocha de Souza for technical support.

CONFLICT OF INTEREST

The authors confirm that this chapter content has no conflict of interest.

REFERENCE

[1]   GUPTA, R.; RAMNANI, P. Microbial keratinases and their prospective applications: an overview. Appl Microbiol Biotechnol, 2006, 70, 21-33.

[2]   RAJPUT, R.; SHARMA, R.; GUPTA, R. Biochemical Characterization of a Thiol-Activated, Oxidation Stable Keratinase from *Bacillus pumilus* KS12. Enzyme Research, 2010, doi:10.4061/2010/132148.

[3]   ANBU, P.; GOPINATH, S. C. B.; HILDA, A.; LAKSHMIPRIYA, T.; ANNADURAI, G. Optimization of extracellular keratinase production by poultry farm isolate *Scopulariopsis brevicaulis*. Bioresource Technology, 2007, 98 (6), 1298-1303.

[4]   MAZOTTO, A. M.; MELO, A. C. N.; MACRAE, A.; ROSADO, A. S.; PEIXOTO, R.; CEDROLA, S. M. L.; COURI, S.; ZINGALI, R. B.; VILLA, A. L. V.; RABINOVITCH, L.; CHAVES, J. Q.; VERMELHO, A. B. Biodegradation of feather waste by extracellular keratinases and gelatinases from *Bacillus* spp. World J Microbiol Biotechnol, 2011, 27 (6), 1355-136.

[5]   BARONE, J. R.; SCHMIDT, W. F.; LIEBNER, C. F. E. Compounding and molding of polyethylene composites reinforced with keratin feather fiber I. Com Sci Technol, 2005, 65, 683–692.

[6]   VERMELHO, A. B.; MAZOTTO, A. M.; NOGUEIRA DE MELO, A. C.; VIEIRA, F. H. C.; DUARTE, T. R.; MACRAE, A.; NISHIKAWA, M. M.; BON, E. P. S. Identification of a *Candida parapsilosis* strain producing extra cellular serine peptidase with keratinolytic activity. Mycopathologia, 2010, 169, 57–65.

[7]   WANG, J.; GREENHUT, W. B.; SHIH, J. C. H. Development of an asporogenic *Bacillus licheniformis* for the production of keratinase. Journal of Applied Microbiology, 2005, 98, 761-767.

[8]   YUE, X. Y.; ZHANG, B.; JIANG, D. D.; LIU, Y. J.; NIU, T. G. Separation and purification of a keratinase as pesticide against root-knot nematodes. World Journal of Microbiology and Biotechnology, 2011, 27 (9), 2147-2153.

[9]   THYS, R. C. S.; BRANDELLI, A. Purification and properties of a keratinolytic metallopeptidase from *Microbacterium* sp. J Appl Microbiol, 2006, 101, 1259–1268.

[10]   SCOTT, J. A.; UNTEREINER, W. A. Determination of keratin degradation by fungi using keratin azure. Med Mycol, 2004, 42: 239-46.

[11]   WAWRZKIEWICZ, K.; WOLSKI, T.; LOBARZEWSKI, J. Screening the keratinolytic activity of dermatophytes *in vitro*. Mycopathologia, 1991, 114, 1-8

[12]   DOZIE, I. N. S.; OKEKE, C. N.; UNAEZE, N. C. A. Thermostable, alkaline-active, keratinolytic proteinase from *Chrysosporium keratinophilum*. World J. Microbiol Biotechnol, 1994, 10, 563-567.

[13]   RAMNANI, P.; GUPTA, R. Optimization of medium composition for keratinase production on feather by *Bacillus licheniformis* RGI using statistical methods involving response surface methodology. Biotechnology Applied Biochemistry, 2004, 40, 191-196.

[14]   HEUSSEN, C.; DOWDLE, E. B. Electrophoretic analysis of plasminogen activators in polyacrylamide gels containing sodium dodecyl sulphate and copolymerized substrates. Anal Biochem, 1980, 102, 196-202.

[15]   VERMELHO, A. B.; ALMEIDA, F. V. S.; BRONZATO, L. S.; BRANQUINHA, M. H. Extracellular metalloproteinases in *Phytomonas serpens*. Can J Microbiol, 2003, 49, 221–224.

[16]   NOGUEIRA DE MELO, A. C.; D'AVILA-LEVY, C. M.; DIAS, F. A.; ARMADA, J. L.; SILVA, H. D.; LOPES, A. H.; SANTOS, A. L.; BRANQUINHA, M. H.; VERMELHO, A. B. Peptidases and gp63-like proteins in *Herpetomonas megaseliae*: possible involvement in the adhesion to the invertebrate host. Int J Parasitol, 2006, 36, 415-22.

[17]   NOGUEIRA DE MELO, A. C.; RIBEIRO, M. D.; SOUZA, E. P.; MACRAE, A.; FRACALLANZA, S. E. L.; VERMELHO, A. B. Peptidase profiles from non-albicans Candida spp isolated from the blood of a patient with chronic myeloid leukemia and another with sickle cell disease. FEMS Yeast Research, 2007, 7, 1004-1012.

[18]   HIBINO, T. Purification and characterization of keratin hydrolase in psoriatic epidermis: application of keratin-agarose plate and keratin-polyacrylamide enzymography methods. Anal Biochem, 1985, 147, 342-352.

[19]   BERNAL, C.; CAIRÓ, J.; COELLO, N. Purification and characterization of a novel exocellular keratinase form *Kocuria rosea*. Enz Microb Technol, 2006, 38, 49-54.

[20]   KOJIMA, M.; KANAI, M.; TOMINAGA, M.; KITAZUME, S.; INOUE, A.; HORIKOSHI, K. Isolation and characterization of a feather-degrading enzyme from *Bacillus pseudofirmus* FA30-01. Extremophiles, 2006, 10, 229-235.

[21]   WAWRZKIEWICZ, K.; LOBAREWSKI, J.; WOLSKI, T. Intracellular keratinase of *Trichophyton gallinae*. Journal of Medical and Veterinary Mycology, 1987, 25, 261-268.

[22]   LOPES, B. G.; SANTOS, A. L.; BEZERRA, C. D.; WANKE, B.; DOS SANTOS, L. M.; NISHIKAWA, M. M.; MAZOTTO, A. M.; KUSSUMI, V. M.; HAIDO, R. M.; VERMELHO, A. B. A 25-kDa Serine Peptidase with Keratinolytic activity secreted by *Coccidioides immitis*. Mycopathol, 2008, 166, 35–40.

[23]   MAZOTTO, A.M.; CEDROLA, S.M.L.; LINS, U.; ROSADO, A.S.; SILVA, K.T.; CHAVES, J.Q.; RABINOVITCH, L.; ZINGALI, R.B.; VERMELHO, A.B. Keratinolytic activity of *Bacillus subtilis* AMR using human hair. Letters in Applied Microbiology, 2010 50 (1), 89-96,

[24]   DUARTE, T. R.; OLIVEIRA, S. S.; MACRAE, A.; CEDROLA, S. M. L.; MAZOTTO, A. M.; SOUZA, E. P.; MELO, A. C. N.; VERMELHO, A. B. Increased expression of keratinase and other peptidases by *Candida parapsilosis* mutants. Braz J Med Biol Res, 2011, 44 (3), 212-216

[25]   WAINWRIGHT, M. A. A new method for determining the microbial degradation of keratin in soil. Experientia,1982, 38, 243-4.

[26]   LETOURNEAU, F.; SOUSSOTTE, V.; BRESSOLLIER, P.; BRANLAND, P.; VERNEUIL, B. Keratinolytic activity of *Streptomyces* sp. S.K1-02: a new isolated strain. Letters in Applied Microbiology, 1998, 26, 77-80.

[27]   LIN, X.; LEE, C. G.; CASALE, E. S.; SHIN, J. C. H. Purification and characterization of a keratinase from a feather-degrading *Bacillus licheniformis* PWD-1. Appl Environ Microbiol, 1992, 58, 3271-3275.

[28]   SYED, D. G.; LEE, J.C.; LI, W.J.; KIM, C.J.; AGASAR, D. Production, characterization and application of keratinase from *Streptomyces gulbargensis*. Bioresour. Technol., 2009, 100, 1868–1871.

[29]   PARK, G.; SON, H. Keratinolytic activity of *Bacillus megaterium* F7-1, a feather-degrading mesophilic bacterium. Microbiological Research, 2009, 164 (4), 478-485.

[30]   KUNERT, J. Effect of reducing agents on proteolytic and keratinolytic activity of enzymes of *Microsporum gypseum*. Mycoses, 1992, 35, 343-348.

[31]   ARIKAWA, Y.; OZAWA, T.; IWASAKI, I. An improved photometric method for determination of sulfite with pararosaniline and formaldehyde. Bulletin of the Chemical Society of Japan, 1968, 41 (6), 1454-1456.

[32]   YAMAMURA, S.; YASUTAKA, M.; QUAMRUL, H.; YOKOYAMA, K.; TAMIYA, E. Keratin degradation: a cooperative action of two enzymes from *Stenetrophomonas* sp. Biochemical and Biophysical Research Communications, 2002, 294, 1138-1143.

[33]   BOCKLE, B., MULLER, R. Reduction of disulfide bonds by *Streptomyces pactum* during growth on chicken feathers. Appl Environ Microbiol, 1997, 63, 790–792.

[34]   PRAKASH, P.; JAYALAKSHMI, S. K.; SREERAMULU, K. Production of keratinase by free and immobilized cells of *Bacillus halodurans* strain PPKS-2: partial characterization and its application in feather degradation and dehairing of the goat skin. Applied Biochemistry and Biotechnology, 2010, 160 (7), 1909–1920.

[35] RITZ, D.; BECKWITH, J. Roles of thiol-redox pathways in bacteria. Annu. Rev. Microbiol, 2001, 55: 21-48.

[36] MAEDA, K.; HÄGGLUND, P.; FINNIE, C.; SVENSSON, B.; HENRIKSEN, A. Structural Basis for Target Protein Recognition by the Protein Disulfide Reductase Thioredoxin. Structure, 14, 1701–1710, 2006.

[37] FABIANEK, R. A.; HENNECKE, H.; THÖNY-MEYER, L. Periplasmic protein thiol:disulfide oxidoreductases of *Escherichia coli*. FEMS Microbiology Reviews, 2000, 24, 303-316.

[38] ELLMAN, G. L. Tissue sulfhydryl groups *Arch.* Biochem. Biophys., 1959, 82, 70-77. (Original determination)

[39] OU, S.; KWOX, K.C.; WANG, Y.; BAO. H. An improvement method to determine SH and –S-S-group content in soymilk protein. Food Chemistry – Anatylical, Nutritional, and Clinical Methods, 2004, 88: 317-320.

[40] RAMNANI, P.; SINGH, R.; GUPTA, R. Keratinolytic potential of *Bacillus licheniformis* RG1: structural and biochemical mechanism of feather degradation. Canadian Journal of Microbiology, 2005, 51 (3), 191-196.

[41] LAEMMLI, U. K. Cleavage of structural proteins during the assembly of the head of bacteriophage T4. Nature, 1970, 227, 680-685.

[42] NOGUEIRA DE MELO, C. A.; GIOVANNI-DE-SIMONE, S.; BRANQUINHA, M. H.; VERMELHO, A. B. *Crithidia guilhermei*: Purification and Partial Characterization of a 62-kDa extracellular metalloproteinase. Experimental Parasitology, 2001, 97, 1-8.

[43] LEBER, T. M.; BALKWILL, F. R. Zymography: A single-step staining method for quantitation of proteolytic activity on substrate gels. Anal. Biochem., 1997, 249, 24–28.

[44] KIM, S. H.; CHOI, N. S.; LEE, W. Y. Fibrin zymography: a direct analysis of fibrinolytic enzymes on gels. Anal Biochem, 1998, 263, 115–116.

[45] FEITOSA, L., GREMSKI, W.; VEIGA SS, ELIAS MC, GRANER E, MANGILI OC, BRENTANI RR. Detection and characterization of metalloproteinases with gelatinolytic, fibronectinolytic and fibrinogenolytic activities in brown spider (*Loxosceles intermedia*). Venom. Toxicon, 1998, 36, 1039-1051.

[46] JAIN, A.; NANCHAHAL, J.; TROEBERG, L.; GREEN, P.; BRENNAN, F. Production of cytokines, vascular endothelial growth factor, matrix metallopeptidases, and tissue inhibitor of metallopeptidases 1 by tenosynovium demonstrates its potential for tendon destruction in rheumatoid arthritis. Arthritis Rheum., 2001, 44, 1754-1760.

[47] ITOH Y.; ITO, A.; IWATA, K., TANZAWA, K.; MORI, Y.; NAGASE, H. Plasma membrane-bound tissue inhibitor of metallopeptidases (TIMP)-2 specifically inhibits matrix metallopeptidase 2 (gelatinase A) activated on the cell surface. J. Biol. Chem., 1998, 273, 24360-24367.

[48] ZENG, Z. S.; SHU, W. P.; COHEN, A. M.; GUILLEM, J. G. Matrix metallopeptidase-7 expression in colorectal cancer liver metastases: Evidence for involvement of MMP-7 activation in human cancer metastases. Clin. Cancer Res., 2002, 8, 144-148.

[49] CEDROLA, S. M. L.; MELO, A. C. N.; MAZOTTO, A. M.; LINS, U. G. C.; ZINGALI, R.B.; ROSADO, A. S.; PEIXOTO, R. S.; VERMELHO, A. Keratinases and sulfide from *Bacillus subtilis* SLC to recycle feather waste. World J Microbiol Biotechnol., 2012, DOI 10.1007/s11274-011-0930-0.

[50] TROEBERG, L.; NAGASE, H. (2003) Current Protocols in Protein Science. 21.15.1-21.15.12 Copyright © 2003 by John Wiley & Sons, Inc.

[51]   VERMELHO, A. B.; MEIRELLES, M. N. L.; LOPES, A; PETINATE, S. D. G.; CHAIA, A. A.; BRANQUINHA, M. H. (1996) Detection of Extracellular Proteases from Microorganisms on Agar Plates. Mem Inst Oswaldo Cruz 91(6): 755-760.

Methods to Determine Enzymatic Activity, 2013, 262-280

CHAPTER 11

Qualitative and Quantitative to Determine Peptidase Activity

Alane Beatriz Vermelho[1,2,*], **Edilma Paraguai de Souza**[1], **Daniel Paiva**[1] and **Giseli Capaci Rodrigues**[1]

[1]*Department of General Microbiology, Institute of Microbiology Paulo de Góes* and [2]*Biotechnology Center-Bioinovar, Federal University of Rio de Janeiro, Rio de Janeiro, Brazil*

Abstract: The aim of this chapter is to present the modern analytical methodologies used to study peptidases, such as the colorimetric methods and spectrophotometric analyses. A general method for the detection of peptidases activity includes gelatin and other substrates incorporated into SDS-PAGE using a technique known as zymography. All methodologies presented here are important tools for the discovery and characterization of peptidases.

Keywords: Peptidase, protease, gelatin, metallopeptidases, isopropanol, zymography.

INTRODUCTION

Peptidases or proteases (EC 3.4) are enzymes that are ubiquitous in nature and are found in all living organisms, including prokaryotes, fungi, plants and animals. They play an essential role in cell growth and differentiation [1]. The extracellular peptidases are of industrial interest and have multiple applications in the food, leather, detergent, and pharmaceutical industries, as well as in diagnostic waste management, among other areas [2, 3]. From among all sources of proteolytic enzymes, microorganisms are the preferred source for industrial applications due to their technical and economic advantages [4].

Gelatinases, which are peptidases that degrade gelatin [5], are one of the most studied enzymes. Several forms of gelatinases are expressed in numerous

*Address correspondence to Alane Beatriz Vermelho: Institute of Microbiology Paulo de Góes & Biotechnology Center BIOINOVAR, at Federal University of Rio de Janeiro, Tel: (021)25626743; Fax: (021) 25608344; E-mail: abvermelho@micro.ufrj.br

microorganisms and play essential roles for their survival, such as in bacteria [6-7], fungi [8] and parasites such as Leishmania [9] and *Trypanosoma cruz*i [10].

Important gelatinases in humans are the matrix metallopeptidases (MMP). The most important of these are MMP-2 and MMP-9 [11-13]. Uncontrolled MMP activity is associated with several pathological processes, such as carcinogenesis.These enzymes play a fundamental role in tumor growth, the multistep processes of invasion and metastasis, as well as proteolytic degradation of the extracellular matrix (ECM), alteration of the cell–cell and cell–ECM interactions, migration and angiogenesis [14].

Thus, the detection of gelatinolytic activity has also become a major ally in the discovery of new gelatinases and therefore new chemotherapeutic targets [15].

Measurement of Gelatinolytic Activity

Quantitative Methods: Gelatinolytic Activity Using Spectrophotometry

Principle of the Method

Methods for detection of gelatinolytic activity have become an important tool for the discovery and characterization of gelatinases. This work provides a simple methodology for the detection of these hydrolytic enzymes incorporating gelatin as substrate and using the Lowry method to determination total protein and adapted by the Jones method [22]. Spectrophotometry is one commonly used method to determine total protein

Gelatin substrate is employed to detect and quantify gelatinolytic activity. Among the existing methods to determine total protein, there is the Lowry method [16] originally proposed by Wu in 1922 [17]. The principle of this method is based on copper-catalyzed oxidation of peptides and proteins by phosphomolybdic /phosphotungstic acid – Folin phenol reagent [18]. The main advantage of the Lowry method is its high sensitivity and, therefore, it has been used to determine the total protein concentration in various media, such as: cerebrospinal fluid (CSF), blood plasma, human saliva, animal tissue, plants, bile juice, membranes, human milk and food products [19, 20]. On the other hand, although the Lowry method has good sensitivity for proteins, it has some disadvantages such as being subject to many types of interference, long analysis time, highly variable specific absorptivity for different proteins, and following the Beer Lambert law only a small range of protein concentration [21]. The original method [16] used trichloroacetic acid as the precipitating agent, but since gelatin is soluble in this acid it is not possible to carried out assays in the classic manner. Therefore the Lowry method [16] was adapted to the Jones method [22] and uses isopropanol as the precipitating agent.

This method used has been shown to be efficient for the characterization of gelatinolytic activity in several microorganisms. Also, this method was comprehensive for gelatinase activity evaluation, since the gelatin is degraded in a nonspecific manner. The experiments should be carried out in triplicate.

Reagents

1. Enzyme preparation (*e.g.,* culture supernatant);

2. Gelatin;

3. Isopropanol;

4. Sodium carbonate (Na_2CO_3);

5. Sodium hydroxide (NaOH);

6. Copper II sulfate pentahydrate ($CuSO_4.5H_2O$);

7. Potassium sodium tartrate ($KNaC_4H_4O_6.2H_2O$);

8. Folin's reagent (sodium 1,2-naphthoquinone-4-sulfonate).

Materials

1. 10mL-glass test tubes;

2. Pipettes (200 µL and 1000 µL);

3. Glass flask (100 mL);

4.   1mL quartz cuvette (1mL);

Equipment

1.   Centrifuge

2.   Spectrophotometer;

3.   Vortex mixer;

4.   Chronometer;

5.   Laboratory oven.

Solutions

1. Na_2CO_3 2% in NaOH 0.1N (Recommendation: only use freshly prepared solutions);

2. $CuSO_4$. $5H_2O$ 1% in distilled water;

3. $KNaC_4H_4O_6$. $4H_2O$ 2% in distilled water;

4.   Folin reagent 50% in distilled water;.

5.   100 mL of solution 1 – 1 mL of solution 2 – 1 mL of solution 3;

6.   Phosphate buffer pH 7.4;

7. Na_2HPO_4 0.06M KH_2PO_4 0.04M (80:20 v/v);

8.   Gelatin 1% in distilled water.

Procedure

Sample Preparation [22]

In a glass test tube, add 100 µL of the sample + 400 µL of phosphate buffer pH 7.4 (the sample relationship verses buffer may vary depending on the sample);

Add 1.5 mL of gelatin 1% solution;

Sample: incubate the test tube for 30 minutes at 37°C;

Blank: remove 750 µL of reaction mixture and add to a tube containing 1 mL of isopropanol. This will be the reaction zero time. To ensure that it will not react store it in the freezer;

After the incubation time, remove 750 µL of the reaction mixture and add it to a tube containing 1 mL of isopropanol;

Incubate for 15 minutes at 4°C;

Centrifuge both (the blank and sample procedure) at 600g for 15 min.

Sample Reading [16]

Use 100 µL of reaction mixture + 400 µL of phosphate buffer pH 7.4;

Add 2.5 mL of 5 solution;

Incubate for 10 min at room temperature;

Add 250 µL of Folin reagent (solution 4) and shake with the aid of a vortex;

Incubate for 30 min at room temperature;

Read using a spectrophotometer at 660 nm;

A standard curve containing bovine serum albumin (BSA) at concentrations of 0, 2.5, 5, 25, 100 and 250 µg (Table **1**) of protein for 1000 µL. To all concentrations of BSA add 2.5 mL of solution 5 and 250 µL of folin reagent.

Table 1: Calibration curve readings carried out by the Lowry method.

BSA (µg)	x*
0	0.070
2.5	0.081
5	0.092

Table 1: contd....

25	0.197
100	0.501
250	0.935

* Mean of triplicate.readings (A_{660} nm)

Triplicates mean were standard curve, Fig. **1**, the x values were calculated using the blank and sample means from the linear equation (Table **2**).

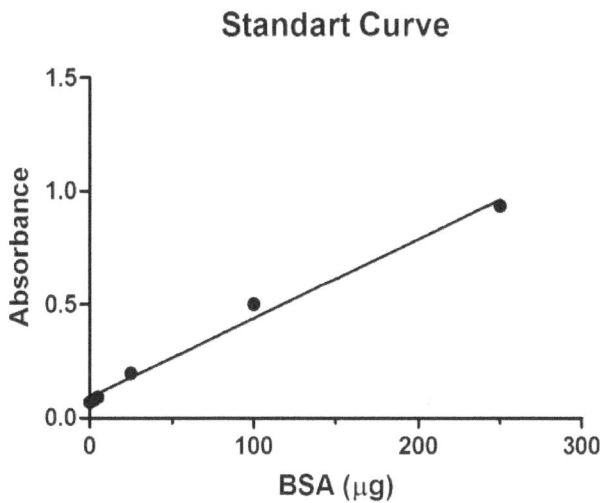

Figure 1: Standard curve obtained with bovine serum albumin y = 0. 0035x + 0. 0906), R^2 = 0.99.

Table 2: Example of the sample and blank reading.

Sample	Blank Reading		Sample Reading		Xtest	GA[a]
	$x^{(b)}$	$x_h^{(c)}$	$x^{(b)}$	$x_s^{(d)}$	$x_s - x_h$	U/mL
1	0.259	48.11	0.296	58.69	10.57	105.71
2	0.267	50.40	0.346	72.97	22.57	225.71
3	0.402	88.97	0.685	169.83	80.86	808.57

(a) Gelatinolytic activity
(b) Mean of triplicates absorbance reading (A_{660} nm)
(c) x_b = blank concentration
(d) x_s = sample concentration

Calculation

Assemble standard curve and obtain the line equation on the chart;

Calculate the x values using the blank and sample means;

$$x= \frac{(mean-0.09060)}{0.0035}$$

Subtract the blank x value from the sample x value;

$$xs - xb$$

1 enzyme unit corresponds to the increase of 1 µg in the readings;

Adjust the volume from 100 µL to 1 mL;

#valor -------------100 µL

X---------------1000 µL

X=valor.10 (U/mL)

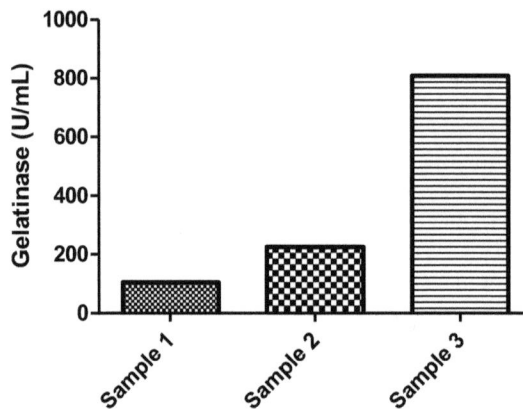

Figure 2: Gelatinolytic activity.

Qualitative Methods: Detection of Extracellular Peptidases

Zymography

Principle of the Method

Zymography is a two-stage versatile electrophoretic technique based on sodium dodecyl sulfate – polyacrylamide gel electrophoresis (SDS-PAGE) which contains a substrate copolymerized within the polyacrylamide matrix involving protein separation.

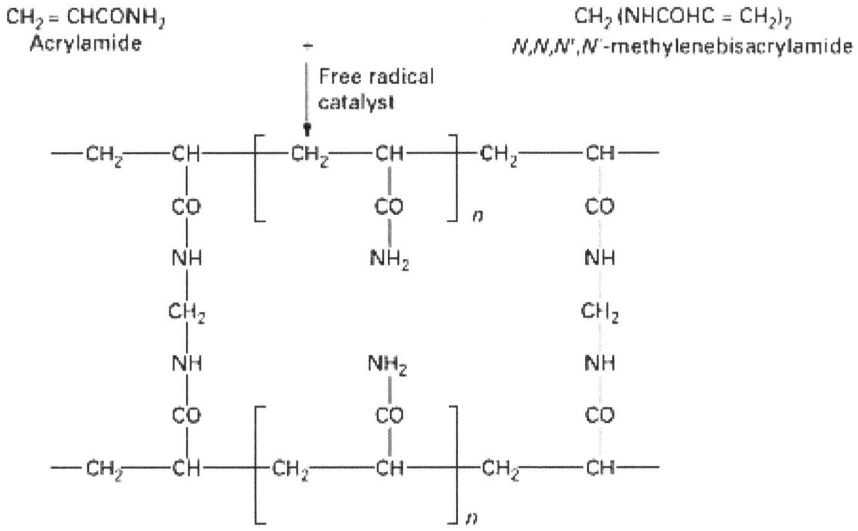

CH$_2$ = CHCONH$_2$
Acrylamide

CH$_2$ (NHCOHC = CH$_2$)$_2$
N,N,N',N'-methylenebisacrylamide

Free radical
catalyst

Figure 3: The formation of polyacrylamide gel from acrylamide and bis-acrylamide.

After separation of the proteins enzyme activity detection is carried out under non-reducing conditions (no treatment of reducing agents such as dithiothreitol or β-mercaptoethanol) [23, 24]. Several substrates can be incorporated in the gel such as casein [25], hemoglobin [26], gelatin [10], collagen [27] and bovine serum albumin [26].

After the electrophoretic run, the SDS is removed of the gel for enzyme renaturation by incubation in a non-buffered Triton X-100 (or similar detergent) and the gels are incubated in an appropriate activation buffer accordingly to the enzyme type being assayed and the substrate being used. Afterwards, the zymogram is stained with Coomassie *Brilliant* Blue and the peptidase activity in visualized as clear or lightly stained bands relative to a darkly stained background indicating where the peptidase has hydrolyzed the substrate and the resulting peptides have diffused out of the gel [23, 28-29].

Biotechnologically, zymography is a simple and powerful technique offering several advantages when studying enzymes. Among these advantages it can:

assess a repertoire of enzymes that have a particular activity in crude cell extracts, estimate the molecular weight and isoelectric point of the enzymes and determine different enzymes with similar and overlapping substrate specificities [28, 30].

Despite their analytical utility, SDS-PAGE based techniques possess a fundamental limitation. This limitation is the lack of some general means for identifying a specific polypeptide of interest from the polypeptide population that can be visualized using dyes such coomassie brilliant blue, silver deposition, [35S] methionine labeling or other proteins "stains [28].

Materials

1. Acrylamide;

2. Bisacrylamide (N,N₁ methylene bisacrylamide);

3. Trizma Base;

4. Proteic substrate (BSA, casein, hemoglobin, gelatin or keratin);

5. Glycerol;

6. Ammonium persulfate;

7. Tetramethylethylenediamine (TEMED);

8. Distilled water;

9. Sodium dodecyl sulfate (SDS);

10. Bromophenol blue;

11. Sample to degrade proteic substrate (culture supernatants);

12. Molecular mass standards (SDS-PAGE standards);

13. Triton X-100;

14. Coomassie blue;

15. Methanol;

16. Acetic acid;

17. Electrophoresis apparatus and comb;

18. Gel-loading pipette tips or Hamilton syringe;

19. Pipettes and flasks;

Equipements

1. Constant voltage power supply;

2. Sealed plastic container.

Reagents and Solutions

Acrylamide / Bisacrylamide Solution (29.9% /0.9%)

Acrylamide-75 g;

Bisacrylamide -2.6 g.

The reagents were dissolved in 200 mL of distilled water and filtered twice. The solution can be stored up to 6 months at 4°C.

Separation Gel Buffer (1.5M Tris pH 8.0)

18.15 g trizma base is dissolved in ~90 mL distilled water. Adjust pH to 8.8 with 6N HCl and adjust volume to 100 mL with distilled water. The solution can be stored up to 1 month at 4°C.

Substrates

1 g of substrates such as BSA, hemoglobin, gelatin are dissolved in 10 mL distilled water. Casein is dissolved in 10 mL Tris 1.5 M pH 8.8 (separation gel buffer).

10% (w/v) Ammonium Persulfate

1 g of ammonium persulfate is dissolved in 10 mL of distilled water. Make fresh solution

10% Sodium Dodecyl Sulfate (SDS)

10 g SDS is dissolved in 100 mL distilled water (solution can be stored up to 3 months at room temperature).

Stacking Gel Buffer (0.5M Tris-HCl pH 6.8)

6 g trizma base is dissolved in ~90 mL distilled water, adjust pH to 6.8 (solution can be stored up to 1 month at 4°C).

Running Buffer (4x)

3 g trizma base, 14 g glycine and 1 g SDS are dissolved in ~900 mL distilled water. Adjust volume to 1000 mL with distilled water. (solution can be stored up to 1 month at 4°C). Dilute upto 4x with distilled water as required.

Sample Loading Buffer

Dissolve 1.2 g SDS and 0.1 g bromophenol blue in 2.5 mL stacking gel buffer and add 7.2 mL glycerol. Adjust volume to 15 mL with distilled water. The solution can be stored up to 1 month at −20°C.

10% Triton X-100

Triton X-100 was used for SDS remotion in order to renature the peptidases.

To prepare the solution add 100 mL Triton X-100 gradually to 700 mL distilled water under gentle agitation. Adjust volume to 1000 mL with distilled water. Store up to 3 months at 4°C. Dilute 4x with distilled water as required.

Staining Solution (Coomassie Blue)

Stock solution (Dissolve 0.5 g Coomassie brilliant blue R-250 in 25 mL distilled water).

Mix 12.5 mL stock solution with 50 mL methanol and 10 mL acetic acid, and adjust volume to 100 mL with distilled water. The solution can be stored up to several months at room temperature.

Distaining (Solution 1)

Mix 500 mL methanol with 100 mL acetic acid and adjust volume to 1000 mL with distilled water. The solution can be stored up to several months at room temperature.

Distaining (Solution 2)

Mix 50 mL methanol with 70 mL acetic acid and adjust volume to 1000 mL with distilled water. The solution can be stored up to several months at room temperature.

Procedures

1. Assemble the gel electrophoresis apparatus accordingly to the manufacturer's instructions;

2. The acrylamide percentage depends on the molecular weight of the peptidases to be visualized. A lower acrylamide percentage gives the gel large sized pores and a higher acrylamide percentage gives the gel small sized pores;

3. Prepare the separation and stacking gels by mixing the solutions according to Tables **1** and **2**, add the ammonium persulfate and the TEMED when the gel is ready to be polymerized:

Table 1: Separation gel (for 10 mL)

Solution / Reagents	7.5%	10%	12.5%	15%
Separation gel buffer (pH8.8)	2.5 mL	2.5 mL	2.5 mL	2.5 mL
Acrylamide / bisacrylamide solution	2.5 mL	3.3 mL	4.2 mL	5 mL
Glycerol	1 mL	1 mL	1 mL	1 mL
10% (w/v) ammonium persulfate (APS)*	100 μL	100 μL	100 μL	100 μL
TEMED*	16 μL	16 μL	16 μL	16 μL
Protein substrate 1%	1.25 mL	1.25 mL	1.25 mL	1.25 mL
10% SDS	100 μL	100 μL	100 μL	100 μL
Distilled Water	2.534 mL	1.734 mL	0.834 mL	0.034 mL

Table 2: Stacking gel

Solution / Reagents	
Acrylamide / bisacrylamide solution	830 µL
Stacking gel buffer	1.25 mL
Distilled water	2.82 mL
10% SDS	50 µL
10% (w/v) ammonium persulfate	60 µL
TEMED	20 µL

The gel is polymerized in a gel caster. First, pour the separation gel and add a thin layer of distilled water, allowing the gel to polymerize and causing the top of the separation gel to form a smooth surface. After the polymerization, discard the distilled water, pour the stacking gel and carefully place a comb to create the wells. Allow the stacking gel to polymerize for 15 to 20 min at room temperature;

Set up the electrophoresis apparatus with the buffer covering the gel and the positive and negative electrodes;

Blend the samples with sample running buffer. Carefully load molecular mass standards and samples into the appropriate wells using syringe or pipette avoiding spillage between slots;

Hook up the apparatus to a power source under appropriate running conditions to separate the protein bands. For example, a 12.5% acrylamide gel should take ~120 min for the running buffer to reach the bottom of the gel;

After the electrophoresis run, disassemble the gel apparatus and separate the stacking gel from the separation gel. Wash the separation gel twice for 15 minutes with SDS remotion and enzyme renaturing solution (10% Triton X-100);

Discard SDS remotion and enzyme renaturing solution (10% Triton X-100) and incubate the separation gel with an appropriate incubation buffer in an airtight container to prevent the solution from evaporating. Incubation time and temperature depends on which protease is being studied;

Next, discard the incubation buffer and add **Staining solution (coomassie blue)** for about 1h at room temperature. DISCARD staining solution (coomassie blue) and add **Distaining solution 1** until clear bands are visible against the dark blue background. After that, discard Distaining solution 1 and store the gel in **Distaining solution 2** until further analysis.

Example

The *Bacillus subtillis* strain AMR was cultivated on phosphate buffered medium (0.06 M $Na_2HPO_4.7H_2O$ and 0.04 M KH_2PO_4, pH 8.0) supplemented with 1% native feather and 0.01% yeast extract for 5 days at room temperature on a rotary shaker (300 rpm). After incubation, the medium was removed by centrifuged at 2,000g (20 min at 4°C) and the supernatant solution concentrated 25-fold by dialysis (cut-off 9 kDa) against PEG 4000 (overnight at 4°C). The concentrated supernatant was used as enzyme extract to analyze peptidases.

Enzyme extract (30 µg protein/4 µL) and sample loading buffer (6 µL) were mixed, and then using a micro-syringe was carefully injected into the gel (10%) containing the proteic substrate. Electrophoresis was performed under a constant voltage power supply at 170 V. After switching off the current, the gel was carefully removed from the apparatus washed twice with enzyme renaturing solution and was kept in incubation buffer (0.5 M Tris-HCl, pH 7.4) for 48 h at 37 °C. Then the gel was stained with staining solution (about 2 h). The gel was distained with distaining solution 1 until the appearance of degradation bands and kept in distaining solution 2. Fig. **4** shows a zymogram with *Bacillus subtillis* peptidases.

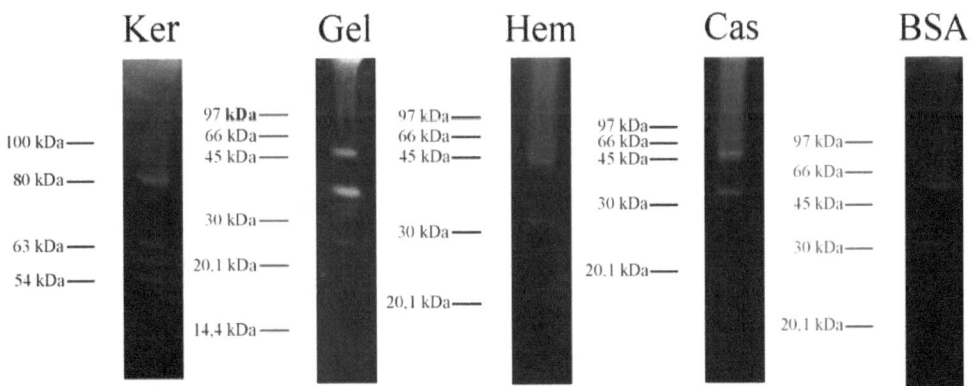

Figure 4: Zymograms with co-polymerized protein substrates. Enzyme extracts of *B. subtilis* strain AMR were prepared as described above. Gel strips were incubated for 48 h at 37 °C in 0.5 M citric acid buffer, pH 5.0. The molecular masses of the peptidases, expressed in kDa, are shown

on the left. The ability of the extracellular peptidases to degrade proteic substrate was evaluated using keratin (Ker), gelatin (Gel), hemoglobin (Hem), casein (Cas) and bovine serum albumin (BSA) co-polymerized with 10% sodium dodecyl sulfate–polyacrylamide gels (SDS–PAGE).

Detection of Extracellular Peptidases on Agar Plates

Principle of the Methods

Detection of extracellular peptidases on agar plates is a valuable technique due to its ability to identify peptidase activity directly in the culture medium. Also, this method allows the use of a broad range of proteic substrates such as BSA, hemoglobin, casein, keratin and gelatin [31, 32]

Materials

1. Agar;

2. Distilled water;

3. Proteic substrate (BSA, casein, hemoglobin, gelatin or keratin);

4. Bromophenol blue;

5. Coomassie blue;

6. Methanol;

7. Acetic acid;

8. Pipettes and flasks;

9. Petri plates;

10. Durhan tube;

11. Inoculating loop.

Reagents and Solutions

Gelatin Agar

Mix.01% (w/v) yeast extract, 2% (w/v) agar and 2% (w/v) gelatin in distilled water (100 mL). This solution is autoclaved, cooled to approximately 50°C and then 20 mL of the suspension is poured gently into a Petri dish and left to harden. Circular wells are dug on the agar plates using sterile Durhan tubes. Now any enzyme extract can be deposited into the wells by dropper.

Nutrient Medium

Choose the appropriate nutrient medium depending on which microorganism will be used.

Saline Solution

0.85% (w/v) NaCl in distilled water.

Staining Solution (Coomassie Blue):

Stock solution (dissolve 0.5 g Coomassie brilliant blue R-250 in 25 mL distilled water).

Mix 12.5 mL stock solution with 50 mL methanol and 10 mL acetic acid, and adjust the volume to 100 mL with distilled water. Store up to several months at room temperature.

Distaining Solution 1

Mix 500 mL methanol with 100 mL acetic acid and adjust volume to 1000 mL with distilled water. Store up to several months at room temperature.

Procedure and Example

The microorganism employed in this study was *Bacillus subtilis* strain AMR isolated from a local poultry industry.

The microorganism was grown in appropriate nutrient medium (2% KCl, 2% sucrose, 0.5% peptone, 0.5% yeast extract and 2% agar in distilled water) for 48h at room temperature. Then a loopful of the culture growth was placed in sterile saline solution and 500 µL of this mixture was placed in the wells of the detection media. After inoculation, the plate was incubated at 37°C and observed daily for five days. Extracellular peptidases detection was made after staining with staining solution (Coomassie blue) overnight. In some cases it may be necessary to use Distaining solution 1.

Regions of enzyme activity were detected as clear areas, indicating that hydrolysis of the substrates had occurred.

Figure 5: Gelatinolytic activity on agar plates containing gelatin as substrate.

ACKNOWLEDGEMENTS

The authors wish to acknowledge the Brazilian National Council for Scientific and Technological Development (CNPq), Carlos Chagas Filho Foundation for Research Support in the State of Rio de Janeiro (FAPERJ), Coordination for the Improvement of Higher Education Personnel (CAPES). The authors are grateful to Denise da Rocha de Souza for technical support.

CONFLICT OF INTEREST

The authors confirm that this chapter content has no conflict of interest.

REFERENCES

[1]   GUPTA, R.; BEG, Q. K.; LORENZ, P. Bacterial alkaline peptidases: molecular approaches and industrial applications. Appl Microbiol Biotechnol., 2002, 59(1), 15-32.
[2]   RAO, M. B.; TANKSALE, A.M.; GHATGE, M. S.; DESHPANDE, V. V. Molecular and biotechnological aspects of microbial peptidases. Microbiol Mol Biol Rev., 1998, 62(3), 597-635.
[3]   SARAN, S.; ISAR, J.; SAXENA, R. K. A modified method for the detection of microbial peptidases on agar plates using tannic acid. J Biochem Biophys Methods. 2007, 10; 70(4), 697-699.

[4]   LAXMAN, R. S.; SONAWANE, A. P.; MORE, S. V.; RAO, B. S.; RELE, M. V.; JOGDAND, V. V.; DESHPANDE, V. V.; RAO, M. B. Optimization and scale up of production of alkaline peptidase from *Conidiobolus coronatus*. Process Biochem. 2005, 40, 3152–3158.

[5]   MURI, E. M.; NIETO, M. J.; SINDELA, R. D.; WILLIAMSON, J. S. Hydroxamic Acids as Pharmacological Agents. Curr Med Chem. 2002, 17, 1631-1653.

[6]   THURLOW, L. R.; THOMAS, V. C.; NARAYANAN, S.; OLSON, S.; FLEMING, S. D.; HANCOCK, L. E. Gelatinase contributes to the pathogenesis of endocarditis caused by *Enterococcus faecalis*. Infect Immun 2010, 78(11): 4936-4943.

[7]   MADATHIPARAMBIL, M. G.; CATTAVARAYANE, S.; PERUMANA, S. R.; MANICKAM, G. D.; SEHGAL, S. C. Presence of 46 kDa Gelatinase on the Outer Membrane of Leptospira. Curr. Microbiol. 2011, [Epub ahead of print].

[8]   DUARTE, T. R.; OLIVEIRA, S. S.; MACRAE, A.; CEDROLA, S.M.; MAZOTTO, A. M.; SOUZA, E. P.; MELO, A. C.; VERMELHO, A. B. Increased expression of keratinase and other peptidases by *Candida parapsilosis* mutants. Braz J Med Biol Res. 2011, 44(3), 212-216.

[9]   MARETTI-MIRA, A. C.; DE OLIVEIRA-NETO, M. P.; DA CRUZ, A. M.; DE OLIVEIRA, M. P.; CRAFT, N.; PIRMEZ, C. Therapeutic failure in American cutaneous leishmaniasis is associated with gelatinase activity and cytokine expression. Clin Exp Immunol. 2011, 163(2), 207-214.

[10]   NOGUEIRA DE MELO, A. C.; DE SOUZA, E. P.; ELIAS, C. G.; DOS SANTOS, A. L.; BRANQUINHA, M. H.; D'AVILA LEVY, C. M.; DOS REIS, F. C.; COSTA, T. F.; LIMA, A. P.; DE SOUZA PEREIRA, M. C.; MEIRELLES, M. N.; VERMELHO, A. B. Detection of matrix metallopeptidase-9-like proteins in *Trypanosoma cruzi*. Exp Parasitol. 2010, 125(3), 256-263.

[11]   MASSOVA, I.; KOTRA, L. P.; FRIDMAN, R.; MOBASHERY, S. Matrix metalloproteinases: structures, evolution, and diversification. FASEB J. 1998, 12(12), 1075-1095.

[12]   CHAKRABORTI, S.; MANDAL, M.; DAS, S.; MANDAL, A.; CHAKRABORTI, T. Regulation of matrix metalloproteinases: an overview. Mol Cell Biochem. 2003, 253(1-2), 269-285.

[13]   HADLER-OLSEN, E.; FADNES, B.; SYLTE, I.; UHLIN-HANSEN, L.; WINBERG, J. O. Regulation of matrix metalloproteinase activity in health and disease. FEBS J, 2011, 278(1), 28-45.

[14]   GIALELI, C.; THEOCHARIS, A. D.; KARAMANOS, N. K. Roles of matrix metalloproteinases in cancer progression and their pharmacological targeting. FEBS J. 2011, 278(1), 16-27.

[15]   DORMÁN, G.; CSEH, S.; HAJDÚ, I.; BARNA, L.; KÓNYA, D.; KUPAI, K.; KOVÁCS, L.; FERDINANDY, P. Matrix metalloproteinase inhibitors: a critical appraisal of design principles and proposed therapeutic utility. Drugs. 2010, 70(8), 949-964.

[16]   LOWRY, O. II.; ROSEBROGII, N. J.; FARR, A. L.; RANDALL, R. J. Protein measurement with the Folin phenol reagent. J Biol Chem. 1951, 193(1), 265-275.

[17]   WU, H. A New Colorimetric Method for the Determination of Plasma Proteins. J Biol Chem. 1922, 51, 33-39.

[18]   LEGLER, G.; MÜLLER-PLATZ, C. M.; MENTGES-HETTKAMP, M.; PFLIEGER, G.; JÜLICH, E. On the chemical basis of the Lowry protein determination. Anal Biochem. 1985, 150(2), 278-287.

[19]   UPRETI, G. C.; RATCLIFF, R. A.; RICHES, P. C. Protein estimation in tissues containing high levels of lipid: modifications to Lowry's method of protein determination. Anal Biochem. 1988, 168(2), 421-427.

[20]   WATERBORG, J. H.; MATTHEWS, H. R. The Lowry method for protein quantitation. Methods Mol Biol. 1994, 32, 1-4.

[21]   SHAKIR, F. K.; AUDILET, D.; DRAKE, A. J.; SHAKIR, K. M. A rapid protein determination by modification of the Lowry procedure. Anal Biochem. 1994, 216(1), 232-233.

[22]   JONES, B. L.; FONTANINI, D.; JARVINEN, M.; PEKKARINEN, A. Simplified endoproteinase assays using gelatin or azogelatin. Anal Biochem. 1998, 263(2), 214-220.

[23]   WILKESMAN, J.; KURZ, L. Peptidase analysis by zymography: a review on techniques and patents. Recent Pat Biotechnol. 2009, 3(3),175-184.

[24]   CHUNG, D.M.; KIM, K. E.; AHN, K. H.; PARK, C. S.; KIM, D.H.; KOH, H. B.; CHUN, H. K.; YOON, B. D.; KIM, H. J.; KIM, M. S.; CHOI, N.S. Silver-stained fibrin zymography: separation of peptidases and activity detection using a single substrate-containing gel. Biotechnol Lett., 2011, Apr 13. [Epub ahead of print]

[25]   MURPHY, G.; SEGAIN, J. P.; O'SHEA, M.; COCKETT, M.; IOANNOU, C.; LEFEBVRE, O.; CHAMBON, P.; BASSET, P. The 28-kDa N-terminal domain of mouse stromelysin-3 has the general properties of a weak metalloproteinase. J. Biol. Chem. 1993, 268,15435-15441.

[26]   MAZOTTO, A. M.; MELO, A. C. N.; MACRAE, A.; ROSADO, A. S.; PEIXOTO, R.; CEDROLA, S. M. L.; COURI, S.; ZINGALI, R. B.; VILLA, A. L. V.; RABINOVITCH, L.; CHAVES, J. Q.; VERMELHO, A. B. Biodegradation of feather waste by extracellular keratinases and gelatinases from *Bacillus* spp. World J Microb. Biot. 2011, 1-11.

[27]   GOGLY, B.; GROULT, N.; HORNEBECK, W.; GO-DEAU, G.; PELLAT, B. Collagen zymography as a sensitive and specific technique for the determination of subpicogram levels of interstitial collagenase. Anal. Biochem. 1998, 255, 211-216

[28]   BISCHOFF, K. M.; SHI, L.; KENNELLY, P. J. The detection of enzyme activity following sodium dodecyl sulfate-polyacrylamide gel electrophoresis. Anal Biochem. 1998, 15;260(1), 1-17.

[29]   PAN, D.; HILL, A. P.; KASHOU, A.; WILSON, K. A.; TAN-WILSON, A. Electrophoretic transfer protein zymography. Anal Biochem. 2011, 411(2), 277-283.

[30]   FREDERIKS, W. M.; MOOK, O. R. Metabolic mapping of proteinase activity with emphasis on *in situ* zymography of gelatinases: review and protocols. J Histochem Cytochem. 2004, 52(6),711-722.

[31]   VERMELHO, A. B.; MEIRELLES, M. N. L., LOPES, A.; PETINATE, S. D. G.; CHAIA, A. A.; BRANQUINHA, M. H. Detection of Extracellular Peptidases from Microorganisms on Agar Plates. Mem Inst Oswaldo Cruz, 1996, 91(6), 755-760.

[32]   LANTZ, M. S.; CIBOROWSKI; P. Zymographic techniques for detection and characterization of microbial peptidases. Methods Enzymol, 1994, 235,563-94.

Send Orders for Reprints to reprints@benthamscience.net

CHAPTER 12

Tannase Activity

Ariana F. Melo and Andrea M. Salgado*

Departament of Biochemical Engineering, Laboratory of Biological Sensors, School of Chemistry, Federal University of Rio de Janeiro, Rio de Janeiro, Brazil

Abstract: The enzyme tannase (E.C.3.1.1.20), also known as tannin acyl hydrolase, is a hydrolytic enzyme that acts on tannins. Tannase is an inducible enzyme that catalyses the breakdown of ester linkages in hydrolysable tannins such as tannic acid resulting in the production of gallic acid and glucose.Most of this enzyme has been to produce gallic acid, which is used in the production of trimethoprim and synthesis of esters such as propyl gallate, an antioxidant used in food industry. The enzyme is also applied in the processing of beer and clarification of juices, instantaneous teas processing and treatment of wastewater contaminated with phenolic compounds. The tannase activity determination is extremely important in several areas, mainly in the food industry analysis.

Keywords: Tannase, tannin hydrolase, tannin, gallic esters, gallic acid, food industry, antioxidant juices, teas, beer, pharmaceutical, methylgallate, ethylgallate, n-propylgallate, isoamylgallate, antimalarial drug, propyl gallate.

INTRODUCTION

The enzyme tannase (E.C.3.1.1.20), also known as tannin acyl hydrolase catalyses the hydrolysis reaction of the ester bonds present in the gallotannins, complex tannins, and gallic acid esters. Tannase is recently commercializedby Biocon (India), Kikkoman (Japan) ASA Specilaeznyme GmbH (Germany) and JFC GmbH (Germany) with different catalytic units depending of the product presentation [1, 2].

Tannase production and application have been extensively studied. These studies related to strain isolation and improvement, process development and application of tannases have conducted to a great number of scientific publications and

---

*Address correspondence to Andrea M. Salgado: School of Chemistry at Federal University of Rio de Janeiro, Brazil; Tel: 55 (21) 2562-7648; Fax: 55 (21) 2562-7567; Email: andrea@eq.ufrj.br.

patents. Table **1** presents some of the published patents regarding tannase production and application.

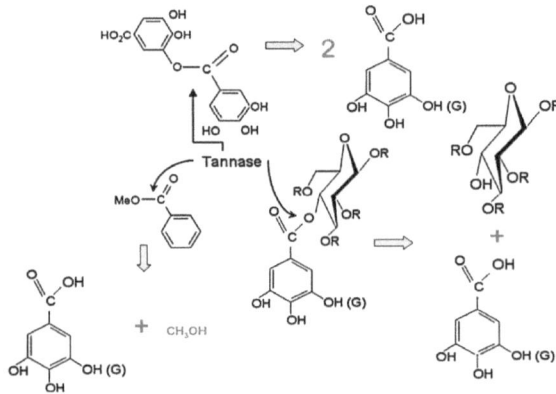

Figura 1: Mechanism of action of tannase. Tannase catalyses the breakdown of hydrolyzable tannins such as tannic acid, methylgallate, ethylgallate, n-propylgallate, and isoamylgallate [3].

Table 1: Application of tannases [6]

Years	Titles	Patent No
1974	Conversion of green tea and natural tea leaves using tannase	US 3812266
1975	Tea soluble in cold water	UK 1280135
1976	Extraction of tea in cold water	GP 2610533
1976	Enzymatic solubilization of tea cream	USP 3953497
1985	Gallic acid ester(s) preparation	EP 137601
1985	Preparation of gallic esters, eg., propylgallate	EP 137601
1985	Enzymatic treatment of black tea leaves	EP 135222
1989	Preparation of spray-concrete coating ming shaft	SUP 1514947
1989	Antioxidant Catechin and gallic acid preparation	JP 01268683
1995	Enzymatic clarification of tea extracts	USP 5445836
2000	Tea concentrates prepared by enzymatic extraction and containing xanthan gum is stable at ambient temperature	USP 6024991
2000	Producing theaflavin	USP 6113965
2006	Diagnostic agent and test method for colon cancer using tannses as index	USP 7090997
2006	Isolation of a dimmer di-gallate a potent endothelium-dependent vasorelaxig compound	USP 7132446
2008	A non-tea-based, packaged beverage with a green tea extract	USP 11845356
2008	A process for preparing a theaflavin-enhaced tea product	USP 11998613
2009	A packaged effervescent beverage having a purified product of green tea extract	USP 123073000

Tannase is widely used in food, feed, beverage, brewing, pharmaceutical and chemical industries. The major applications of this enzyme are in the production of gallic acid, which is used for the manufacture of an antimalarial drug, trimethoprim and in the synthesis of esters as propyl gallate used as antioxidants in the food industry[4,]. The enzyme is also used in the manufacture of instant tea, in the prevention of phenol-induced mediarization in wine, manufacture of coffee-flavoured soft drinks, clarification of beer and fruit juices, stabilization of malt polyphenol and as a sensitive analytical probe for determining the structure of naturally occurring gallic acid esters [5,6]. Tannin hydrolysis with tannase at lower temperatures and under gentle conditions, can reduce flavor loss. Cream is hydrolyzed to lower molecular weight components, reducing turbidity and increasing cold water solubility, hence soluble tannase has been recommended for use in the preparation of instant tea. Enzyme solubilization of cream is optimal at 45°C with a rate inversely proportional to tea solids concentration. However, the enzyme must be subsequently denatured by heating to over 90°C, resulting in a loss of tea flavor.

The enzyme is used in food and beverage processing; however, the practical use of this enzyme is at present, limited due to insufficient knowledge about its properties, optimal expression, and large-scale application.

A number of methods such as tritriometric [7] (spectrophotometric [8, 9,], UV spectrophotometric, gas chromatographic, and high-performance liquid chromate-graphy assays have been developed to study tannase activity.

Methods for tannase assay based on the titration of gallic acid released by the action of the enzyme on tannic acid did not give correct results due to the problem of accurately determining the end point. Other method for tannase assay based on the estimation of glucose liberated by incubation with the enzyme for 24 h are reported, which is not suitable for routine assays of the enzyme [9].

Each method has its own limitations and most are time consuming and require instrument sophistication. Considering this, this chapter provides an overview of the colorimetric assay methods for determination of activity tannase [10-12].

METHOD 1

A Spectrophotometric Method for Assay of Tannase Using Rhodanine

Introduction

Tannase catalyzes the hydrolysis of ester and depside linkages in hydrolyzable tannins like tannic acid. The products of hydrolysis are glucose and gallic acid. Most of the methods for tannase assay based on the titration of gallic acid released by the action of the enzyme on tannic acid did not give correct results due to the problem of accurately determining the end point. A method for assay of microbial tannase (tannin acyl hydrolase) based on the formation of chromogen between gallic acid and rhodanine, unlike the previous protocols, this method is sensitive up to gallic acid concentration of 100mg/L. The assay is complete in a short time, very convenient, and reproducible.

Objectives

This method is used to determine microbial tannase activity by spectrophotometry at 520 nm. In fast and precise basis.

Test Principle

A colorimetric assay based on the formation of chomogen between gallic acid and rhodanine.

Materials and Methods

Reagents and solutions required:

Reagents

1. Tannic acid;

2. Rhodanine;

3. Gallic acid;

4. Potassium hydroxide;

5. Ethanol;

6. Acetic acid;

7. Sodium acetate.

All other chemicals were of an analytical grade.

Solution

1. Tannic acid solution in acetate buffer 20mM pH 5,0 (0,02% w/v);

2. Ethanolicrhodanine solution (0.667% w/v);

3. Potassium hydroxide solution (0.5 M);

4. Gallic acid solution (100 mg/L).

Procedure

Calibration Curve for Gallic Acid Estimation

Gallic acid (100mg/L) in distilled water was used as standard. Aliquots of this solution containing 10 – 100 mg/L gallic acid were taken, and the volume was made to 0,4 mL with water. The blank contained 0,4mL distilled water. 0,6 mL rhodanineethanolic solution was added to all the tubes, and they were incubated for 5 min. After 0,4 mL potassium hydroxide solution was added to all the tubes which were again incubated for 2,5 min. Distilled water (8,6 mL) was added to all the, and after 20 min absorbance was recorded at 520 nm.

For the standard curve, the absorbance is ploted on the Y axis against the concentration of gallic acid (mg/L) in the X axis. From the slope of the fitted straight line, calculate the conversion factor.

Tannase Assay

The reaction mixture containing 0.25 mL enzyme sample, and 10 mL tannic acid (0.02% w/v) was incubated for 10 min at 30°C. Ethanolic rhodanine solution (0.6 mL) was added for stopping the reaction and for formation of complex between gallic acid and rhodanine. The tubes were kept for 5 min. After this, 0.4 mL of 0.5 M potassium hydroxide was added to each tube and these were incubated for 2.5 min. This was followed by addition of the enzyme sample (0.25 mL) to the reaction mixture in the control tube only. Finally, each tube was diluted with 8.6 mL of distilled water and incubated for 10 min and the absorbance was recorded against water at 520 nm using a spectrophotometer:

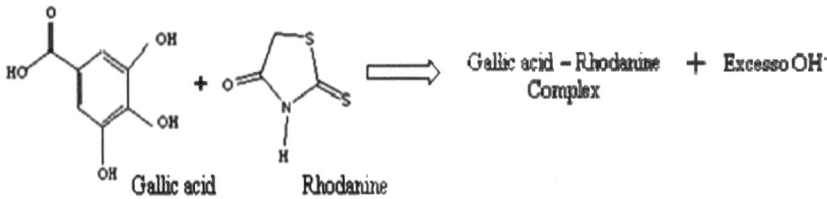

Figure 2: Formation of the complex: Gallic acid with rhodanine.

One unit (U) of enzymatic activity is defined as the amount of enzyme that delivers 1 micromol of gallic acid per minute of reaction in a pH of 5,0 at 30 °C.

The calculation of the free enzyme activity was done by equation 1:

$$ATIV = \frac{\left(Abs_{test} - Abs_{blank}\right) \times Dil \times f \times 21}{170,1 \times 10}$$

(1)

Where:

$ATIV \rightarrow$ Tanase Activity (U/mL)
$Abs_{test} \rightarrow$ Test sample absorbance
$Abs_{blank} \rightarrow$ Blank absorbance
$Dil \rightarrow$ Enzimatic complex dilution
$f \rightarrow$ Conversion factor of the gallic acid standard curve (mg/L)

Results and Conclusions

The proposed method, was utilized to determination of tannase synthesized by solid substrate fermentations [13]. For this purpose *A. niger*3T5B8, 11T25A5 and

11T53A9, selected previously by their potential in tannase synthesis [14] were inoculated in solid state medium. The results showed that *A. niger*3T5B8 was the best producer of this enzyme, with 1,67U.g-1 after 72 hours of fermentation.

Pinto *et al.* [13] has shows that the utilization, of this method for routine analysis of tannase, was confirmed by the results shown in Fig. **3**. An increase in protein concentration, from 20.0 to 161.1 μg/mL of protein, followed by incubation for 10 minutes in a 200 μg/mL solution of tannic acid, gives a linear response of activity. The prime factor for an enzymatic assay is to guarantee that all active enzyme molecules are bound to the substrate, once the product formation is linearly dependent of the complex enzyme-substrate. The method proposed here is capable of to determine tannase activity in a large range of protein concentration

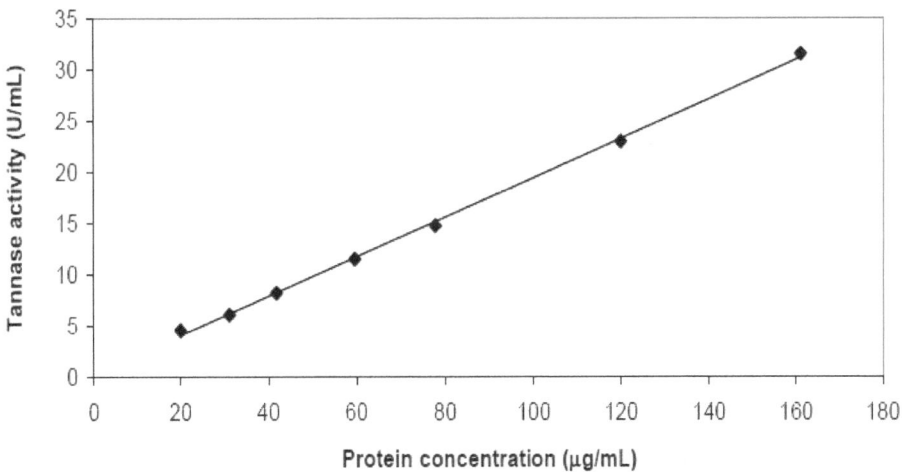

Figure 3: Relation between protein concentration and tannase activity measured by the ethanolic rhodanine method. A linear relation was observed - $R^2 = 0,9989$ [13].

METHOD 2

Colorimetric Tannase Assay Method

Introduction

Tannic acid forms a brown color in alkaline medium with FeCl3. It is closer to a colorimetric filter range of 530 nm. A modified protein precipitation method of

Hagerman and Butler was adopted here to measure the residual tannic acid. Due to the polyphenolic nature of tannic acid it can easily bind and precipitate any protein, such as BSA solution. The salt (NaCl) composition of BSA solution enhances the formation of the tannin–protein complex. The SDS–triethanolamine is an alkaline detergent, which disrupts the tannin–protein complex. The FeCl3 reacts with phenolic groups of tannin by nucleophilic attraction and forms a brown color based on this principle. Mondal *et al.* 15] developed a colorimetric method to quantify tannase activity, which is disbelieved this item.

Objectives

Colorimetric method of tannase assay developed using its specific substrate tannic acid.

Test principle

Tannase activity was determined by the changes in optical density of substrate tannic acid after enzymatic reaction at 530nm.

Materials and Methods

Reagents

1. Commercial tannic acid;

2. Bovine serum albumin;

3. SDS;

4. Triethanolamine;

5. $FeCl_3$ reagent;

6. Acetic acid;

7. Sodium acetate;

8. Hydrochloric acid.

    All reagents are of analytical grade and can be stored for long time, except tannic acid.

Solution

1. Tannic acid solution 0.5% (w/v) in 0.2 M acetate buffer (pH 5.5) was used as substrate;

2. Bovine serum albumin solution 1 mg/mL was prepared with 0.17 M sodium chloride in 0.2 M acetate buffer pH 5.0;

3. SDS–triethanolamine solution. 1% (w/v) solution containing 5% (v/v) of triethanolamine was prepared in distilled water;

4. $FeCl_3$ reagent was prepared with 0.01 M $FeCl_3$ in 0.01 M hydrochloric acid

Procedure

The reaction mixture consisting of substrate tannic acid 0.3 mL and 0.05 mL enzyme was incubated at 60°C for 10 min. The enzymatic reaction was stopped by the addition of 3 mL BSA solution, which also precipitated the remaining tannic acid. In the same way a reference tube was prepared with heat denatured enzyme. The tubes were centrifuged (5000*g* during 5 min) and the resultant precipitate was dissolved in 3 mL SDS–triethanolamine solution. Then 1 mL of FeCl3 reagent was added and kept for 15 min for stabilization of the color. The absorbencies of both the tubes were measured at 530 nm, against the blank, without tannic acid.

Unit of tannase activity can be defined as the amount of enzyme which is able to hydrolyze 1 mM of substrate tannic acid in 1 min under assay conditions.

Results and Conclusions

A unit of enzyme activity can be easily calculated at any time from the optical density values with respect to substrate utilization.

In this assay system a maximum of 1.5 mg of tannic acid (0.05 mL of 3% w/v or 0.3 mL of 0.5% w/v) may be present in the fermentation broth, being within the range suitable substrate for enzymatic reaction. Above this concentration (3% w/v) tannic acid in the fermented broth is necessary to standardize the amount of precipitation of BSA for the better precipitation.

Principal Required Laboratory Equipment for Tests

A Spectrophotometric Method for Assay of Tannase Using Rhodanine

- Spectrophotometer;

- Analytical Electronic Balance;

- Incubator;

- pH meter;

- Water distiller system.

Colorimetric Tannase Assay Method

- Spectrophotometer;

- Centrifuge;

- Heating bath;

- Analytical Electronic Balance;

- pH meter;

- Water distiller system.

ACKNOWLEDGEMENTS

This work was supported by the Brazilian National Council for Scientific and Technological Development (CNPq), Carlos Chagas Filho Foundation for Research Support in the State of Rio de Janeiro (FAPERJ).

CONFLICT OF INTEREST

The authors confirm that this chapter content has no conflict of interest.

REFERENCES

[1]   BARTHOMEUF, C.; REGERAT, F.; POURRAT, H. Production, purification and characterization of a tannase from *Aspergillus niger* LCF 8. J. Ferment. Bioeng. 1994; 77, 320–323.

[2]   BELMARES, R.; CONTRERAS-ESQUIVEL, J.C.; RODRÍGUEZ-HERRERA, R; RAMÍREZ CORONEL, A.; AGUILAR, C.N. Microbial production of tannase: na enzyme with potential use in food industry. Lebensm Wiss Technol 2004; 37:857–864

[3]   AGUILAR, C. N.; RODRÍGUEZ, R.; SÁNCHEZ, G. G.; AUGUR,C.; TORRES,E. F.; PRADO-BARRAGAN, L A.; RAMÍREZ-CORONEL, A.; CONTRERAS-ESQUIVEL, J. C. Microbial sources.Advances. Perspectives Appl. Microbiol Biotechnol 2007; 76:47–59

[4]   AGUILAR, C.N.; GUITIÉRREZ-SÁNCHEZ, G. Review: sources, properties, applications and potential uses of tannin aclyhydrolase. Food Sci. Tech. Int, 2001; v. 5, p. 373-382.

[5]   LEKHA, P. K.; and LONSANE, B. K. Production and application of tannin acyl hydrolase: State of the art. Adv. Appl. Microbiol. 1997; 44, 215–260

[6]   BELUR, P. D.; and MUGERAYA, G.; Microbial Prodution of tannase: State of Art. Journal of Microbiology 2011; 6 (1) 25-40.

[7]   NISHIRA, H. Studies on tannin decomposing enzyme of molds.Tannase fermentation by molds in liquid culture with phenolic substances. Journal Ferment. Technology. 1961; 39, 137–146

[8]   .HASLAM, E.; and TANNER, J. N. Spectrophotometric assay of tannase. Phytochemistry 1970; 90, 2305–2309.

[9]   SHARMA, S.; BHAT, T.K.; DAWRA, R.K. A Spectrophotometric Method For Assay Of Tannase Using Rhodanine. Analytical Biochemistry. 2000; v. 279, p. 85-89.

[10]   IIBUCHI, S.; MINODA, Y.; YAMADA, K. Studies on tanninacyl hydrolase. Part II. A new method determining the enzyme activity using the change of ultraviolet absorption.Agric. Biol.Chem. 1967; 31, 513–518.

[11]   HAGARMAN, A. E.; AND BUTLER, L. G. Protein precipitation method for the quantitative determination of tannins. J. Agric. Food. Chem. 1976; 26, 809–812.

[12]   JEAN, D.; POURRAT, H.; POURRAT, A.; CARNET, A. Assay oftannase (tannin acyl hydrolase EC 3.1.1.20) by gas chromatography. Anal. Biochem. 1981; 110, 369–372.

[13]   PINTO, G.A.S.; COURI, S.; GONÇALVES, E.B. Replacement of methanol by ethanol on gallic acid determination by rhodanine and its impacts on the tannase assay. EJEAFChe, 2006; ISSN: 1579-4377.

[14]   PINTO, G. A. S.;LEITE, S.G.F.; TERZI, S C.; COURI, S. Selection of tannase produccing Aspergillus niger. Brazilian Journal of Microbiology, São Paulo, 2001; v. 32, n. 1, p. 24-26, 2001

[15]   MONDAL, K.C.; BANERJEE, D.; PATI, B.R Colorimetric Assay Method for Determination of the Tannin Acyl Hydrolase (EC 3.1.1.20) Activity. Analytical Biochemistry 2001; (295), 168-171.

Send Orders for Reprints to reprints@benthamscience.net

CHAPTER 13

Urease Activity

Lívia Maria da C. Silva[1], Andrea M. Salgado[1,*] and Maria Alice Z. Coelho[2]

[1]*Departament of Biochemical Engineering, Laboratory of Biological Sensors, School of Chemistry, Federal University of Rio de Janeiro, Rio de Janeiro, Brazil and* [2]*Departament of Biochemical Engineering, Biological Systems Engineering, School of Chemistry, Federal University of Rio de Janeiro, Rio de Janeiro, Brazil*

Abstract: Urease, a nickel-dependent metalloenzyme, is synthesized by a wide variety of organisms, including plants, bacteria and fungi. It catalyzes the hydrolysis of urea into ammonia and carbon dioxide. Although the amino acid sequences of plant and bacterial ureases are closely related, some biological activities differ significantly. To date, the structural information is available only for bacterial ureases although the enzyme extracted from jack bean (*Canavalia ensiformis*), which is the best-studied plant urease, had been crystallized in 1926. Tests for urease activity determination should be based on direct or indirect methods and most of the methods utilized for urease detection are based on analysis of its activity in urea hydrolysis. The urease activity determination is extremely important in several areas, mainly in agriculture, medicine and clinical analysis

Keywords: Urease, metalloenzyme, urea, ammonia, carbon dioxide, canavalia ensiformis, virulence factors, colorimetric test, bromcresol purple, ficoll reagent, nessler, alcalimetric test, pH changes, back-titration, agriculture, clinical test, phenol red, pH indicator, spectrophotometry.

INTRODUCTION

James B. Sumner crystallized, in 1926 [1], the enzyme urease from jack bean, *Canavalia ensiformis* (Fabaceae), a bushy annual tropical american legume grown mainly for forage, demonstrating that enzymes can be crystallized [1]. Urease is abundant enzyme in plants and, moreover, it can be found in numerous eukaryotic microorganisms and also bacteria. The bacterial and plant ureases have high similarity in amino acid sequence, suggesting that they have similar three-dimensional structures and a conserved catalytic mechanism.

*Address correspondence to Andrea M. Salgado:** School of Chemistry at Federal University of Rio de Janeiro, Brazil; Tel: 55 (21) 2562-7648; Fax: 55 (21) 2562-7567; E-mail: andrea@eq.ufrj.br.

Ureases are cysteine-rich enzymes and functionally belong to the superfamily of amidohydrolases and phosphotriesterases. The primary common feature of these enzymes is the presence of a metal centre in the active site, whose task is to activate the substrate and water for the reaction. Among other dinuclear metallohydrolases in the superfamily, ureases are unique since possess Ni(II) ion in the active site, event discovered in 1978, opening a new field of bioinorganic chemistry of nickel.

Ureases (urea amidohydrolase, EC 3.5.1.5) catalyzes the hydrolysis of urea to yield ammonia (NH$_3$) and carbamat, the latter compound decomposes spontausly to generate a second molecule of ammonia and carbon dioxide (CO$_2$) [2] (Fig. **1**). This enzyme also catalyzes the hydrolysis of some urea-substituted and esters of carbamic acid, such as thiourea and *p*-nitrophenilcarbamate, but hydroxyurea and dihydroxiurea are the specific substrates for urease. Different compounds are urease inhibitors, such as acetohydroxamic acid and *p*-hydroxymercuribenzoate [3,4].

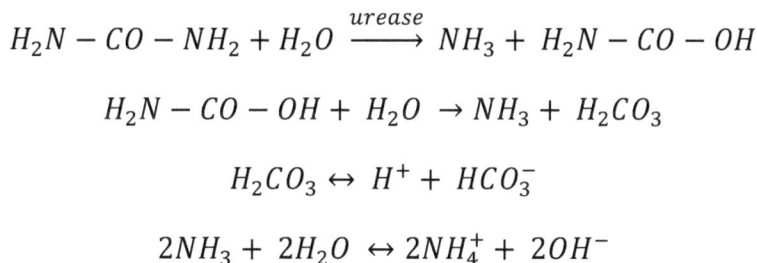

$$H_2N - CO - NH_2 + H_2O \xrightarrow{urease} NH_3 + H_2N - CO - OH$$

$$H_2N - CO - OH + H_2O \rightarrow NH_3 + H_2CO_3$$

$$H_2CO_3 \leftrightarrow H^+ + HCO_3^-$$

$$2NH_3 + 2H_2O \leftrightarrow 2NH_4^+ + 2OH^-$$

Figure 1: Urea hydrolysis catalyzed by urease.

Even today, more than 80 years after Sumner's discovery, a catalytic mechanism of urease has not been fully resolved, and the enzyme, due to its multiple implications, is still extensively studied. The knowledge on the urease active site has been provided by the crystal structures resolved for ureases from two bacteria, *Klebsiella aerogenes* and *Bacillus pasteurii*. While plant and fungal ureases are known to mostly be homohexamers, bacterial ureases typically are heterotrimers, whose units exhibit high homology of amino-acid sequences with one subunit of jack bean urease. Importantly, in all known ureases, the active sites are always located in the same subunits [5].

Urease activity determination is extremely important in several areas. Medically, bacterial ureases are important virulence factors implicated in the pathogenesis of many clinical conditions such as pyelonephritis, hepatic coma, peptic ulceration and the formation of infection-induced urinary stones. This feature is a basic diagnostic criterion used in the determination of many bacteria species, which produce highly active urease. Ureolytic activity of bacteria such as *Clostridium perfringens*, *Klebsiella pneumoniae*, *Proteus mirabilis*, *Salmonella sp.*, *Staphylococcus saprophyticus*, *Ureaplasma urealyticum*, and *Yersinia enterocolitica* plays an important role in the pathogenesis of human and animal diseases.

One of the most frequently studied bacterial urease is that from *Helicobacter pylori* since it has been implicated in peptic ulcers and stomach cancer. The presence of urease is therefore used in the diagnosis of *Helicobacter* species. Moreover, urea - an end product of nitrogen metabolism - has great significance in clinical chemistry [6].

Moreover, determination of blood urea nitrogen is an important routine test widely used in clinical laboratories, as high urea level in blood sera deciphers kidney disease, stone in urinary tract or even bladder tumour. Whereas, lower levels indicate severe liver bad function. The normal range of urea in human serum is between 1.7 and 8.3 mM and level increases up to 100 mM under patho-physiological conditions.

In agriculture, urease plays an important role in the efficient use of urea fertilizer in many soils. The highest activity of urease was determined in embryonic plant tissues, like in seeds of *Fabaceae* and *Curcubitaceae* species, and it is suggested to play an important role in seed germination and also participate in seed chemical defense [7,8].

Numerous direct and indirect detection methods for the reaction products have been successfully applied in the quantification of urease activity. The most commonly used are: the indophenol assay for the detection of ammonium, the quantification of $^{14}CO_2$ by scintillation counting after breakdown of radiolabeled urea and coupled enzyme assays mostly utilizing glutamate dehydrogenase, measuring the oxidation of NADPH spectrophotometrically.

Although colorimetric and spectrometric methods are most commonly used, simple and fast method for determination of urea is in demand, like biosensors. Urea biosensors utilize the biospecificity of urease. In other words, analyte (urea) and enzyme (urease) result in a product (ammonium ion) that can be detected and quantified using a transducer (amperometric, potentiometric, optical, thermal or piezoelectric) [9,10].

This, this chapter provides an overview of the methods for urease activity determination in several areas.

UREASE ACTIVITY DETERMINATION: COLORIMETRIC TEST MONITORADED BY SPECTROPHOTOMETRY

METHOD 1

Introduction

Urease can catalyze the hydrolysis of urea to ammonium, which is monitored by the pH-sensitive dye bromcresol purple. Bromcresol purple turns purple at an alkaline pH and turns yellow at an acid pH. So the color of the solution changed from yellow to purple with increasing time, which indicates an increase in pH due to the ammonium production. The absorbance of the solution at 588 nm is plotted against the reaction time. The urease activity was determined from the rate of ammonium production.

Objective

This test is used to monitor the urease activity by spectrophotometry at 588 nm.

Test Principle

A colorimetric assay based on the hydrolysis of urea was used for the urease activity control, as monitored by the pH-sensitive dye bromcresol purple.

Materials and Methods

Those solutions are required:

1. Urea solution (25mM);

2. Bromcresol purple solution (0.015 mM) as indicator;

3. EDTA solution (0.2 mM);

4. Urease solution (0.01 mg/mL).

Procedure

A mixture containing 1.0 mL of urea solution, 0.5 mL of bromcresol purple solution and 1.0 mL of EDTA solution was adjusted to pH 5.8. This solution was placed in a 3 mL UV cell and magnetically stirred at 25°C for 15 minutes. Either 0.10 mL of urease solution was added to the assay solution. The enzymatic reaction was monitored by following the dye absorption at 588 nm. The solution changed from yellow to dark purple during the reaction, corresponding to a change in solution pH from 5.8 to 7.5.

Results and Conclusions

The increment of the absorbance with time was used to characterize the urease activity in the sample. The urease activity was determined from the rate of ammonium production with time.

METHOD 2

Introduction

The indophenol reaction has been successfully applied for the quantification of urease activity and has several advantages: it is sensitive (ammonium can be quantified in dilute solutions of less than 4 mM); it gives a linear response over a wide range of ammonium concentrations (at least up to 160 mM); the color produced is stable; and the method is inexpensive and requires no sophisticated equipment. Amino acids and peptides have been shown to interfere strongly with less toxic modifications of the original method in which phenol is replaced by salicylate and hypochloride by dichloroisocyanurate [11-13].

Objective

This test is used to monitor the urease activity by spectrophotometry at 636 nm.

Test Principle

Urease activity was determined by measuring ammonia concentration by the colorimetric phenol–hypochlorite method in samples withdrawn from the reaction mixture at 5 minutes reaction.

Materials and Methods

Solutions that are required in the colorimetric phenol–hypochlorite method:

Phenolic Solution

Reagents	Quantity (g)
Phenol	7.0
Disodium pentacyanonitrosylferrate	34.0
This reagent was stored in a dark-coloured bottle at 4°C.	

The phenol solution was prepared as: phenol (7 g) and disodium pentacyanonitrosylferrate (34 mg) were dissolved in deionised water (50 mL) and then made up to 100 mL.

Buffered Hypochloride Reagent

Reagents	Quantity
Hydroxide sodium	2.96 g
Disodium phosphate heptahydrate	22.29 g
Sodium hypochlorite	16.6 mL (12% v/v)
Deionised water	
Final pH: 12.0 at 25°C	
This reagent was stored in a dark-coloured bottle at room temperature.	

The buffered hypochloride reagent was prepared by dissolving 2.96 g of hydroxide sodium in 140 mL of deionised water, adding 22.29 g of disodium phosphate heptahydrate and dissolving it completely. Then sodium hypochlorite (12% v/v, 16.6 mL) solution was added. The pH was adjusted to 12.0 and the deionised water was added to complete the final volume of 200 mL.

Procedure

The urease assay mixture consisted of 100 mM urea in 50 mM phosphate buffer, pH 7.8, with 2 mM EDTA, its volume being 25 mL. Reactions were initiated by the addition of small aliquots of the enzyme-containing (0.5 mg) solution, and the activity was determined by measuring ammonia concentration by the colorimetric phenol–hypochlorite method in samples withdrawn from the reaction mixture at 5 minutes reaction. The measurements were performed at room temperature.

Ammonium (colorimetric method) was measured as follows: 40.0 μL of NH_4Cl (range 7.0 μg/mL – 7.0 mg/mL) was pipetted to glassy test tubes. Further, 1960 μL of deionised water, 200 μL of the phenolic solution and 400 μL of buffered hypochloride reagent were added. The mixture was vortexed for 5 minutes and stored for 20 minutes at 50°C. The coloured solutions were measured by spectrophotometer at 636 nm against a blank sample, which contains 2 mL of deionised water, 200 μL of the phenolic solution and 400 μL of the buffered hypochloride reagent. The blank sample was also stored for 20 minutes at 50°C prior to measurements.

Results and Conclusions

The enzyme activity was determined from the rate of ammonium production by the colorimetric phenol–hypochlorite method in 5 minutes.

METHOD 3

Objective

This test is used to monitor the urease activity by spectrophotometry at 480 nm.

Test Principle

Spectrophotometric stop rate determination.

Materials and Methods

Those solutions are required for the urease activity measurement:

Reagent A: Potassium phosphate buffer (10 mM), pH 8.2 at 30°C with 10 mM Lithium chloride and 1 mM Ethylenediaminetetraacetic acid;

Prepare 100 mL in deionized water using potassium phosphate dibasic trihydrate, lithium chloride anhydrous and ethylenediaminetetraacetic acid tetrasodium salt hydrate. Adjust to pH 8.2 at 30°C with 1 M HCl.

Reagent B: Urea solution (66 mM)

Prepare 25 mL of reagent B using urea and reagent A.

Reagent C: Ficoll solution (0.4% (w/v))

Prepare 20 mL in deionized water using Ficoll® reagent.

Reagent D: Nessler's color reagent

Dissolve 100 g mercuric iodide (II) and 70 g potassium iodide in 100 mL deionized water, adding a cold solution of 160 g sodium hydroxide in 700 mL of deionized water and bringing the solution to a volume of 1L. Allow to settle the precipitate for a few days before using the reagent.

Reagent E: Ammonia standard solution (2.50 mM)

Prepare 50 mL in deionized water using ammonium sulfate reagent.

Reagent F: Urease solution

Immediately before use, prepare a solution containing 100-150 units/mL of urease in cold reagent A.

Procedure

Pipette 10 mL of Reagent B into a suitable container, incubate at 30°C, then at time zero add 0.025 mL of Reagent F. Immediately mix by inversion and remove 1.0 mL aliquot from this solution (Test). Transfer to a suitable container containing 5.0 mL of Reagent D and mix by inversion. This is to be used as the Reagent Blank. Continue to incubate at 30°C. Then at 2, 4 and 6 minutes (time increments), transfer 1.0 mL Test solution into suitable containers containing 5.0 mL of Reagent D and mix by inversion. Transfer the solutions to suitable cuvette and record the A480nm for the Reagent Blank and Tests using a spectrophotometer.

Colorimetric of Standards

Reagent	Standards (mL)						
	Blank	1	2	3	4	5	6
Deionized water	1.00	0.90	0.80	0.60	0.40	0.20	0
Reagent E	0	0.10	0.20	0.40	0.60	0.80	1.00
Reagent D	5.00	5.00	5.00	5.00	5.00	5.00	5.00

Mix the reagents by inversion. Transfer to suitable cuvettes, measure and record the A_{480nm} for the Standard Blank and Standards.

Results and Conclusions

To calculate the enzyme activity, first make a standard curve necessary to calculate ΔA_{480nm} Standard (ΔA_{480nm} Standard = A_{480nm} Standard - A_{480nm} Standard Blank) and plot μmoles of NH_3 *versus* ΔA_{480nm} to obtain ΔA_{480nm}/μmole NH_3. After, calculate ΔA_{480nm} Test (ΔA_{480nm} Test = A_{480nm} Test - A_{480nm} Reagent Blank) and ΔA_{480nm} Test/min (ΔA_{480nm} Test/min = ΔA_{480nm} Test/Time increment (in minutes)). The urease activity is calculated using the equation below:

$$\frac{(\Delta A_{480nm}\ Test\ /\ minute)\ (1000)}{(\Delta A_{480nm}/\mu mole\ NH_3)(mg\ enzyme/mL\ RM)} = Units\ /\ g\ enzyme$$

where:

RM = Reaction mixture volume

1000 = Conversion factor for converting milligrams to grams

One unit will liberate 1.0 μmole of ammonia from urea per minute at pH 8.2 at 30°C.

METHOD 4

Introduction

Worthington Biochemical Corporation® has adopted an assay method where the hydrolysis of urea is measured by coupling ammonia production to a glutamate dehydrogenase reaction, as shown in the following Fig. **2**.

$$Urea + H_2O + 2H^+ \xrightarrow{urease} 2NH_4^+ + CO_2 \, 2NH_4^+ + 2\alpha - ketoglutarate$$
$$+ 2NADH \xrightarrow{GLKM} 2 \, glutamate + 2NAD^+ + 2H_2O$$

Figure 2: Hydrolysis of urea with a coupled glutamate dehydrogenase reaction.

Objectives

This test is used to monitor the urease activity by spectrophotometry at 340 nm.

Test Principle

Urease activity is measured as a result of oxidation of NADH at 25°C and pH 7.6 under the specified conditions.

Materials and Methods

Those solutions are required for the urease activity measurement:

1. Potassium phosphate buffer (0.1 M, pH 7.6);

2. Adenosine-5'-diphosphate (ADP) solution (0.023 M) in potassium phosphate buffer;

3. NADH solution (0.0072 M) in potassium phosphate buffer;

4. α-ketoglutarate solution (0.026 M) in potassium phosphate buffer;

5. Urea solution (1.8 M) in potassium phosphate buffer;

6. Glutamate dehydrogenase (GLDH) solution: (dilute to approximately 500 units/mL in 50% glycerol or potassium phosphate buffer; store cold during use);

7.   Urease solution (dissolve enzyme at 1 mg/mL in potassium phosphate buffer; immediately prior to use, dilute further in buffer to obtain a rate of 0.02-0.04 ΔA/minute).

Procedure

First, adjust spectrophotometer to 340 nm. So, place a mixture containing 2.4 mL of potassium phosphate buffer, 0.1 mL of ADP solution, 0.1 mL of NADH solution, 0.1 mL of α-ketoglutarate solution, 0.1 mL of urea solution and 0.1 mL of GLDH solution in a cuvette at 25°C for 5-10 minutes to achieve temperature equilibrium. A slight change in absorbance may be observed due to trace ammonia in reagents. Upon obtaining a zero change in absorbance, add 0.1 mL appropriately diluted enzyme. Record decrease in A_{340} for 8-10 minutes. Determine ΔA_{340}/minute from the linear portion of the curve. A slight lag may occur.

Results and Conclusions

One unit results in the oxidation of one micromole of NADH per minute at 25°C and pH 7.6 under the specified conditions. The urease activity is calculated using the equation below:

$$\frac{\Delta A_{340nm} / min}{6.22 \times mg\ enzyme\ /\ mL\ reaction\ mixture} = Units/mg\ enzyme$$

UREASE ACTIVITY DETERMINATION: ALCALIMETRIC TEST

METHOD 1

Introduction

Alkalimetric method is based on the observation made by Kistiakowsky & Shaw [1953 apud,14] which the initial pH neutral of unbuffered solution of urea-urease rapidly increases to pH 9.0, and then remains approximately constant. The reaction products in this pH are usually ammonium carbamate, ammonium carbonate and bicarbonate as shown in the following Fig. **3**:

$$H_2NCONH_2 + 2H_2O \rightarrow (NH_4)^+ + (NH_2CO_2)^- + H_2O$$

$$\rightarrow 2(NH_4)^+ + (CO_3)^-$$

$$\rightarrow (NH_4)^+ + NH_3 + (HCO_3)^-$$

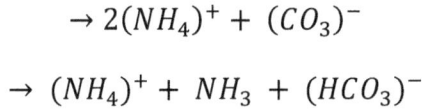

Figure 3: Ammonium carbamate, ammonium carbonate and bicarbonate as urease reaction products.

These end products of the reaction are a buffer system that maintains the pH constant as the reaction proceeds. So, using the substrate initially buffered at pH 9.0, avoids the subsequent change in pH. The addition of excess hydrochloric acid in the final time disrupts the reaction and converts the carbamate and ammonia to ammonium ions. Therefore, back-titration with sodium hydroxide measures the acid did not react [14].

Objective

This test is used to determine the urease activity by alkalimetric method by back-titration with sodium hydroxide.

Materials and Methods

Those solutions are required:

1. Urea solution (3%, w/v) in Tris/HCl buffer solution;

2. Tris/HCl solution (0.1 M, pH 9,0);

3. Urease solution (0.98 mg/mL);

4. Sodium hydroxide solution (0.05 N);

5. Hydrochloric acid solution (0.1 N);

6. Methylorange solution (0.2 g/ 100 mL) as indicator.

Procedure

In this method, the urease activity was assayed by adding 1mL of urea solution, 1mL of urease solution and 10mL of deionised water. Incubation was carried out at 25°C (room temperature) for a constant interval. Withdraw an aliquot (2 mL) of mixture solution and stop the reaction by adding hydrochloric acid solution. Then, the reaction mixture was back-titrated with sodium hydroxide solution, being methylorange used as an indicator. The blank test was assayed under the same conditions above, using 1mL of urea solution and 11 mL of deionised water.

Results and Conclusions

To calculate the enzyme activity, first is necessary to calculate the volume of sodium hydroxide (vol. NaOH) which is given by: vol.NaOH = vol.NaOH blank – vol.NaOH test. So, the urease activity calculated using the equation below:

$$\frac{(NaOH\ molarity)(vol.NaOH)(1000)(df)}{(t)(1.0)} = \ nits/_{mL\ enzyme}$$

where:

Vol.NaOH = Volume (in milliliters) of sodium hydroxide solution used in the back titration

1000 = Conversion factor from millimoles to micromoles

$df\ =\ Dilution\ factor$

$t\ =\ Time\ of\ assay\ (in\ minutes)$

$1.0\ =\ Volume\ (in\ milliliter)\ of\ enzyme\ used$

METHOD 2

Objective

This test is used to monitor the urease activity by titration method.

Test Principle

The measurement of enzyme activity is carried out by titration method using hydrochloric acid as tritant.

Materials and Methods

Reagent A: Sodium Phosphate Buffer (750 mM), pH 7.0 at 25°C

Prepare 200 mL in deionized water using sodium phosphate monobasic anhydrous. Adjust to pH 7.0 at 25°C with 1 M NaOH. Store at room temperature.

Reagent B: Urea solution (500 mM) with 0.05% (w/v) bovine serum albumin solution.

Prepare using 50 mL Reagent A with urea and bovine albumin. Adjust to pH 7.0 at 25°C with 100 mM HCl or 100 mM NaOH, if necessary.

Reagent C: 0.10% (w/v) 3-(4-dimethylamino-1-naphthylazo)-4-methoxybenzenesulfonic acid solution as indicator.

Prepare 100 mL in deionized water using 3-(4-dimethylamino-1-naphthylazo)-4-methoxybenzenesulfonic acid solution. Facilitate dissolution by first adding 2.6 mL of 100 mM NaOH, then adjusting the final volume to 100 mL with deionized water.

Reagent D: Standardized hydrochloric acid solution (100 mM)

Prepare 100 mL in cold deionized water using concentrated hydrochloric acid.

Reagent E: Sodium phosphate buffer (20 mM), pH 7.0 at 25°C as enzyme diluents.

Prepare 100 mL in deionized water using sodium phosphate monobasic anhydrous. Adjust to pH 7.0 at 25°C with 1 M NaOH.

Reagent F: Urease solution

Immediately before use, prepare a solution containing 200-400 units/mL of urease in cold Reagent E.

Procedure

Blank Solution

Pipette 1.0 mL of Reagent B into a suitable container, then add 3.0 mL of Reagent D, 0.1 mL of Reagent F and two drops of Reagent C. Using a magnetic stirrer, titrate immediately with Reagent D by adding small amounts until the color of the indicator turns from orange to pink. This is the endpoint. Record the volume of Reagent D used for Blank solution.

Test Solution

Pipette 1.0 mL of Reagent B into a suitable container, then add 0.1 mL of Reagent F. Mix by stirring and incubate at 25°C for exactly 5 minutes. Then add 3.0 mL of Reagent D, 0.1 mL of Reagent F and two drops of Reagent C. Using a magnetic stirrer, titrate immediately with Reagent D by adding small amounts until the color of the indicator turns from orange to pink. This is the endpoint. Record the volume of Reagent D used for the Test solution.

Results and Conclusions

To calculate the enzyme activity, first is necessary to calculate the volume of hydrochloric acid (vol. HCl) which is given by:

$$vol. HCl = vol. HCl\ test - vol. HCl\ blank$$

So, the urease activity calculated using the equation below:

$$\frac{(HCl\ molarity)(vol. HCl)(1000)(df)}{(5)(0.1)} = Units/mL\ enzyme$$

where:

$$vol. HCl = Volume\ (in\ milliliters)\ of\ Reagent\ D\ used\ in\ the\ titration$$

1000 = Conversion factor from millimoles to micromoles as per the Unit definition

$$df = Dilution\ factor$$

$$5 = Time\ of\ assay\ (in\ minutes)\ as\ per\ the\ Unit\ Definition$$

$$0.1 = Volume\ (in\ milliliter)\ of\ enzyme\ used$$

One unit will liberate 1.0 μmole of ammonia from urea per minute at pH 7.0 at 25°C, under the assay conditions. It is equivalent to 1.0 I.U. or 0.054 Sumner unit (1.0 mg ammonia nitrogen in 5 minutes at pH 7.0 at 20°C).

UREASE ACTIVITY DETERMINATION IN AGRICULTURE

METHOD 1

Introduction

The enzyme urease is present in soybeans, but is of very limited interest in monogastric nutrition. However, much like trypsin inhibitors and lectins, its activity is reduced by heating. As a determination of urease activity is far easier to conduct than are assays for trypsin inhibitor, urease activity is frequently used as a "marker" to indirectly reflect the presence of antinutritional factors in soy products. Historically, urease activities with a 0.15 increase in pH units suggested underprocessing, while activities of less than 0.05 units indicated overprocessing. However, during the past several decades, a change in soybean processing has led to the production of meals with lower changes of 0.05 in pH with no apparent detrimental effect on animal performance. In addition, the former maximum acceptable variation of 0.15 pH units is no longer considered as absolute. Older birds, especially laying hens, can easily tolerate meals with a urease value of 0.25 or possibly even higher. The relative insensitivity of the urease assay as a means of quantifying overprocessing of soybean meal is due to the fact that there is no negative scale in this assay. The test is unable to distinguish between meals which may be barely acceptable *versus* those which are grossly overprocessed.

Objective

Laboratory tests are needed to determine whether samples of soybean meal have received adequate, but not excessive, heat treatment following oil extraction. This enzyme is present in uncooked soybean and is progressively destroyed by heat.

Test Principle

Determines the activity of residual urease in soybean products for use as an indicator of sufficient heat treatment; urease catalyzes urea breakdown and results in a pH rise due to ammonia generation.

Materials and Methods

Solution Preparation

Phosphate Buffer Solution (0.05 M phosphate buffer) (g/L):

Reagents	Quantity
Monobasic potassium phosphate	3.403 g
Dibasic potassium phosphate	4.355 g
Distilled water	
Final pH: 7.0 at 25°C	

Dissolve 3.403 g of monobasic potassium phosphate in approximately 100 mL of distilled water. Dissolve 4.355 g of dibasic potassium phosphate in approximately 100 mL of distilled water. Combine the two solutions and dilute to 1000 mL. Use a stir bar and stir plate to mix solutions. Adjust the pH of the buffer solution to 7.0 with a strong acid or base before using. Use buffer within 90 days.

Buffered Urea Solution

Reagents	Quantity
Urea	15.0 g
Phosphate buffer solution	500 mL
Final pH: 7.0 at 25°C	

Dissolve 15 g of urea in 500 mL of the phosphate buffer solution. Add 5 mL of toluene to serve as a preservative and to prevent mold formation. Use a stir bar and stir plate to mix solutions. Adjust the pH of the urea solution to 7.0 with a strong acid or base before using.

Procedure

Place approximately 0.200 g (± 0.001 g) of sample into a test tube (sample should be finely ground - grind large particles in coffee grinder). Add 10 mL of the buffered urea solution - this will be called the **test sample.** Mix/swirl gently (do not invert tube) and place in water bath at 30°C noting the time put into the bath.

Place approximately 0.200 g (± 0.001 g) of sample into a test tube (sample should be finely ground - grind large particles in coffee grinder). Add 10 mL of the phosphate buffer solution - this will be called the **blank sample**. Mix/swirl gently (do not invert tube) and place in water bath at 30°C noting the time put into the bath.

In both test and blank samples tubes, mix/swirl contents of tubes approximately every five minutes during the required time in the water bath. At end of 30 minutes, remove tubes (each tube will have a different removal time from the water bath depending on time put in) and mix/swirl contents one last time.

Allow tube to stand for a few minutes and transfer approximately 5 mL of the supernatant liquid into a small vial or beaker (vial or beaker must provide for electrode of pH meter to fit into it and to be covered with the liquid).

Determine the pH of the supernatant liquid approximately five minutes after removal from the water bath.

RESULTS AND CONCLUSIONS

Calculate differences between test pH and blank pH as an index of urease activity:

$$Urease\ Index\ (pH\ change) = (pH\ of\ test\ sample) - (pH\ of\ blank\ sample)$$

UREASE ACTIVITY DETERMINATION IN CLINICAL TESTS

METHOD 1

Introduction

Stuart, Van Stratum, and Rustigian (1945) [15] first developed a highly buffered urea broth containing the pH indicator phenol red to determine the presence of urease. This medium made it possible to differentiate *Proteus* species, an active urease producer, from other members of the family *Enterobacteriaceae*. Once the urea has been split and ammonia is formed, an alkaline pH shift occurs, and a pinkish-red color is produced.

Later, Christensen (1946) [16] developed a medium which included peptone and dextrose and in which the buffer content was reduced. This medium is capable in determining urease activity for members of the family *Enterobacteriaceae*, as well as other organisms, which exhibit weak or delayed urease activity. This medium eliminated the need for the organisms to use urea as the sole source of nitrogen, and is more sensitive, demonstrating urease activity in organisms that are more complex in their nitrogen requirement. *Proteus* species may show positive results in 1-6 hours, whereas delayed urease-positive organisms may only exhibit a weakly positive reaction after several days of incubation. The indicator is the same in both media; Christensen's urea turns pinkish-red when urea is split and an alkaline pH is reached.

In 1967, Philpot developed a formula for the differentiation of *Trichophyton mentagrophytes* from *Trichophyton rubrum* by a simple urease test. *Trichophyton mentagrophytes* splits urea more rapidly in urea agar with added dextrose than do other dermatophytes. *Trichophyton rubrum*, except for some granular strains, gives a negative or a weak-positive reaction. The urea agar base is supplied as a 10 times concentrated solution for use in preparing Urea Agar (Christensen) slants in the laboratory.

Objective

Urea media are used to determine the ability of an organism to split urea *in vitro* by the action of the enzyme urease, especially for *Proteus* species.

Principle

Some organisms possess the urease enzyme which is capable to hydrolyze urea yielding ammonia and ammonium carbonate as end products (alkaline products).

Materials and Methods

Solutions that are required in the urea test (per liter deionized filtered water):

Urea Broth

Reagents	Quantity (g)
Yeast extract	0.1
Monopotassium phosphate	9.1
Disodium	9.5
Urea	20.0
Phenol red	0.01
Final pH: 6.8 ± 0.2 at 25°C	

Urea Agar (Christensen's Medium), Tube

Reagents	Quantity (g)
Pancreatic digest of gelatin	1.0
Dextrose	1.0
Sodium chloride	5.0
Monosodium phosphate	2.0
Urea	20.0
Agar	15.0
Phenol red	0.012
Final pH: 6.8 ± 0.2 at 25°C	

Urea Dextrose Agar

Reagents	Quantity (g)
Pancreatic digest of gelatin	1.0
Dextrose	5.0
Sodium chloride	5.0
Monosodium phosphate	2.0
Urea	20.0
Agar	15.0
Phenol red	0.012
Final pH: 6.8 ± 0.2 at 25°C	

Urea Agar Base (10 times concentrate)

Reagents	Quantity (g)
Pancreatic digest of gelatin	10.0
Dextrose	10.0

Sodium chloride	50.0
Monosodium phosphate	20.0
Urea	200.0
Phenol red	0.12
Final pH: 6.8 ± 0.2 at 25°C	

Upon receipt store at range 2-8°C away from direct light. Media should not be used if there are signs of contamination, deterioration or if the expiration date has been passed.

Procedure

Method of Use (tube): Prior to inoculation, the medium should be brought to room temperature. For Urea Broth, the medium is inoculated with a suspension of organisms taken from a young broth culture (*e.g.*, tryptone), or growth from 18 to 24-hour pure culture. Shake the tube gently to suspend the bacteria, loosen cap, and incubate aerobically at 35°C for 24-48 hours. For Urea Agar, inoculate from 18 to 24- hour pure culture with a straight wire. Using a fishtail motion, cover the entire slant, then loosen cap and incubate aerobically at 35°C for up to 6 days. Examine daily for urease activity.

Method of Use (plate): If using a multipoint inoculation system, lightly touch the top of one or two well-isolated and morphologically similar colonies and inoculate into a broth culture. Deposit a broth spot 5-6 mm in diameter onto the surface of the agar plate using the replicator or any comparable device that could be readily adaptable. Incubate aerobically at 35°C for 18-24 hours. Directions for preparation of medium from Urea Agar Base 10 times Concentration: Prepare Urea Agar (Christensen's medium) by allowing 90 mL of an autoclaved 1.5% agar solution to cool to 50°C, and then aseptically add 10 mL of concentrate (the contents of one tube) to each 90 mL of cooled agar solution. Mix thoroughly and dispense aseptically into sterile test tubes. Allow the tubes to cool in a slanted position so that slants with deep butts are formed.

Results and Conclusions

Urea Broth

Urea broth is highly buffered and will therefore mask urease activity for organisms which exhibit a weak or delayed reaction. Therefore, this medium should only be used for the detection of members of the tribe *Proteeae*. For a strong positive reaction, *Morganella morganii* may require 36-48 hours of incubation. Moreover, false-negative results can occur if only small amounts of ammonia are produced, or if caps are tight.

Interpretation	
Results	**Color**
Positive	Intense pink-red color throughout the broth
Negative	No color change (yellow-orange)

Urea Agar (Christensen's medium)

Proteus species will split urea soon after inoculation. Results can be read within the first 1-6 hours of incubation. *Citrobacter freundii* and *Klebsiella pneumoniae* may take 24-48 hours to split urea. Some organisms exhibit a delayed reaction which may occur up to 6 days after inoculation.

Interpretation	
Results	**Color**
Positive	Intense pink-red (red-violet) color
Negative	No color change or change to pale yellow

Urea Dextrose Agar

A positive test after four days is not significant.

Interpretation	
Results	**Color**
Positive	Intense pink-red (red-violet) color
Weak Positive	Pale pink color in media
Negative	No color change or change to pale yellow

Quality Control	
Urea Broth and Urea Agar	
Results	**Species**
Positive	*Proteus mirabilis* (ATCC 12453)
Negative	*Escherichia coli* (ATCC 25922)
Urea Dextrose Agar	
Positive	*Trichophyton mentagrophytes*
Negative	*Trichophyton rubrum*

METHOD 2

Objectives

This urease test is used to differentiate urease-positive *Proteus* species from other members of Enterobacteriaceae, but some strains of *Enterobacter* and *Klebsiella* species are also urease-positive. The test may also be used to distinguish *Psychrobacter phenylpyruvicus* from *Moraxella* species and *Corynebacterium diphtheriae* (Urease-negative) from *C. ulcerans* and *C. pseudotuberculosis* (urease-positive).

Test Principle

The urease test is used to determine the ability of an organism to split urea by production urease enzyme. Two units of ammonia are formed with resulting alkalinity. The production of alkali is detected by a pH indicator (phenol red) presents in Christensen's urea.

Materials and Methods

Christensen's Medium (g/L)

Reagents	Quantity
Pancreatic digest of gelatin	1.0
Dextrose	1.0
Sodium chloride	5.0
Monosodium phosphate	2.0
Urea	20.0
Agar	15.0
Phenol red	0.012
Final pH: 6.8 ± 0.2 at 25°C	

Dissolve 29 g of the medium in 100 mL of distilled water. Sterilize by filtration. Separately dissolve 15 g of agar in 900 mL of distilled water by boiling. Sterilize in autoclave at 121°C for 15 minutes. Cool to 50°C and add to the 100 mL of the sterile Urea Agar Base. Mix well and dispense aseptically in sterile tubes. Leave the medium to set in a slanted position so as to obtain deep butts. Do not remelt the slanted agar. The prepared medium should be stored at 2-8°C. The color at a pH of 6.8 to 7.0 should have a light pinkish-yellow color.

Procedure

All identification tests should be performed from a non-selective medium. Inoculate slope heavily over the entire surface. Incubate slope heavily over the entire surface. Incubate inoculated slope ate 35-37°C in a water bath or hot block or incubator. Examine slopes after 4 hours and after overnight incubation.

Results and Conclusions

In case of positive test, the medium turns pink under alkaline conditions (pH 8.4) due to phenol red indicator in the medium. No change in colour (yellow, pH 6.8) indicates negative reaction (Fig. **4**).

Figure 4: Positive test; Negative test or No-inoculated medium.

Quality Control	
Results	**Species**
Positive	*Providencia rettgeri* (NCTC 7475)
Negative	*Serratia marcescens (*NCTC 11935)

PRINCIPAL REQUIRED LABORATORY EQUIPMENT FOR TESTS

UREASE ACTIVITY DETERMINATION: COLORIMETRIC TEST MONITORADED BYSPECTROPHOTOMETRY

- Spectrophotometer;

- Analytical Electronic Balance;

- Plate shaker;

- Vortex;

- Incubator;

- pH meter;

- Water distiller system.

UREASE ACTIVITY DETERMINATION: ALCALIMETRIC TEST

- Titrator;

- Analytical Electronic Balance;

- Plate shaker;

- Vortex;

- Incubator;

- pH meter;

- Water distiller system.

UREASE ACTIVITY DETERMINATION IN AGRICULTURE

- Water bath;

- Analytical Electronic Balance;

- Plate shaker;

- pH meter;

- Incubator;

- Water distiller system.

UREASE ACTIVITY DETERMINATION IN CLINICAL TESTS

- Analytical Electronic Balance;

- Vortex;

- pH meter;

- Laminar air flow cabinet;

- Incubator;

- Autoclave;

- Water distiller system;

- Water deionizer system.

  Supplementary materials can be found in the following references and URLs: 16-27.

ACKNOWLEDGEMENTS

This work was supportedby the Brazilian National Council for Scientific and Technological Development (CNPq) and Carlos Chagas Filho Foundation for Research Support in the State of Rio de Janeiro (FAPERJ).

CONFLICT OF INTEREST

The authors confirm that this chapter content has no conflict of interest.

REFERENCES

[1] SUMNER, J. B. The isolation and crystallization of the enzyme urease. J. Biol. Chem. 1926; 69, pp. 435-441.

[2] TAKISHIMA, K.; SUGA, T.; MAMIYA, G. The structure of jack bean urease. European Journal of Biochemistry 1988; 175, pp. 151-165.

[3] FOLLMER, C.; REAL-GUERRA, R.; WASSERMAN, G.E; OLIVERA-SEVERO, D.; CARLINI, C.R. Jackbean, soybean and *Bacillus pasteurii* ureases: Biological effects unrelated to ureolytic activity. European Journal of Biochemistry 2004; 271, pp. 1357–1363.

[4] BARROS, T. G.; WILLIAMSON, J. S.; ANTUNES, O. A. C.; MURI, E. M. F. Hydroxamic acids designed as inhibitors of urease. Letters in Drug Design & Discovery 2009; 6 (3), pp. 186-192.

[5] KRAJEWSKA, B.; ZABORSKA, W. Jack bean urease: The effect of active-site binding inhibitors on the reactivity of enzyme thiol groups. Bioorganic Chemistry 35, pp. 355-365.

[6] NECCHI, V.; MANCA, R.; RICCI, V., SOLCIA, E. Evidence for transepithelial dendritic cells in human *H. pylori* active gastritis. Helicobacter 2007; 14 (3), pp. 208-222.

[7] CHAKRABARTI, K.; SINHÁ, N.; CHAKRABORTY, A.; BHATTACHARYYA, P. Influence of soil properties on urease activity under different agro-ecosystems. Archives of Agronomy and Soil Science 2004; 50 (4-5), pp. 477-483.

[8] ASKIN, T.; KIZILKAYA, R. The spatial variability of urease activity of surface agricultural soils within an urban area. Journal of Central European Agriculture 2005; 6 (2), pp. 161-166.

[9] SINGH, M.; VERMA, N.; GARG, A. K.; REDHU, N. Urea biosensors. Sensors and Actuators B 2008; 134, pp. 345–351.

[10] DHAWAN, G.; SUMANA, G.; MALHOTRA, B. D. Recent developments in urea biosensors. Biochemical Engineering Journal 2009; 44, pp. 42–52.

[11] WEATHERBURN, M. W. Phenol-Hypochlorite Reaction for Determination of Ammonia. Analytical Chemistry 1967; 39 (8), pp. 971-974.

[12] CHAKRABARTI, K.; SINHÁ, N.; CHAKRABORTY, A.; BHATTACHARYYA, P. Influence of soil properties on urease activity under different agro-ecosystems. Archives of Agronomy and Soil Science 2004; 50 (4-5), pp. 477-483.

[13] WITTE, C-P.; MEDINAS-ESCOBER, N. In-Gel detection of urease with nitroblue tetrazolium and quantification of the enzyme from different crop plants using the indophenol reaction. Analytical Biochemistry 2001; 290, pp. 102–107.

[14] COMERLATO, M. H. Imobilização de enzimas no suporte crisotila. 90 p. Dissertação (Doutorado em Química) – Universidade Estadual de Campinas, Campinas - SP. 1995.

[15] STUART, C.A., VAN STRATUM, E., a. RUSTIGIAN, R. Further studies on urease production by Proteus and related organisms. J. Bact., 1945; 49; pp437-444

[16] CHRISTENSEN, W.B. Urea decomposition as means of differentiating Proteus and Paracolon cultures from each other and from Salmonella and Shigella types. J. Bact., 1946; 52, pp461-466.

[17] DIXO, N. E.; GAZZOLA, C.; BLAKELEY, R. L.; ZERNER, B. Jack bean urease (EC 3.5.1.5) metalloenzyme. Simple biological role for nickel J. Am. Chem. Soc. 1975; 97 (14), pp 4131–4133.

[18] HUBALEK, J.; HRADECKY, J.; ADAM, V.; KRYSTOFOVA, O.; HUSKA, D.; MASARIK, M.; TRNKOVA, L.; HORNA, A.; KLOSOVA, K.; ADAMEK, M.; ZEHNALEK, J.; KIZEK, R. Spectrometric and voltammetric analysis of urease – nickel nanoelectrode as an electrochemical sensor. Sensors 2007; 7, pp. 1238-1255.

[19] LIANG, Z.; WANG, C.; TONG, Z.; YE, W.; YE, S. Bio-catalytic nanoparticles with urease immobilized in multilayer assembled through layer-by-layer technique. Reactive & Functional Polymers 2005; 63, pp. 85–94.

[20] LVOV, Y.; CARUSO, F. Biocolloids with ordered urease multilayer shells as enzymatic reactors. Anal. Chem. 2001; 73, pp. 4212-4217.

[21] WORTHINGTON, C. E. Worthington Enzyme Manual, 1972; pp. 146-148, Worthington Biochemical Corporation, Freehold, NJ available in http://www.worthington-biochem.com/URC/assay.htmL

[22] BioMerieux's procedure for Urea Media (2009) available in: http://www.pmLmicro.com/assets/TDS/795.pdf

[23] National Standard Method: Urease test available in: http://www.hpa.org.uk/cfi/esl/default.htm

[24] Sigma's procedure for Enzymatic Assay of Urease from Jack Beans (1995) available in: http://www.sigmaaldrich.com/etc/medialib/docs/Sigma/General_Information/urease_from_jack_beans.Par.0001.File.dat/urease_from_jack_beans.pdf

[25] Sigma's procedure for Enzymatic Assay of Urease (1997) available in http://www.sigma-aldrich.com/etc/medialib/docs/Sigma/General_Information/2/urease.Par.0001.File.dat/urease.pdf

[26] Chemistry 39 (8), pp. 971-974

[27] http://www.poultry.uga.edu/soybeans/qualitycontrol.htm

[28] http://www.analytichem.com/Applications/Kjeldahl/urease%20activity%20in%20soybean.htm

Subject Index

www.ingramcontent.com/pod-product-compliance
Lightning Source LLC
Chambersburg PA
CBHW050808220326
41598CB00006B/150